Darwin, Tennyson and Their Readers

Darwin, Tennyson and Their Readers

Explorations in Victorian Literature and Science

Edited by Valerie Purton

ANTHEM PRESS
LONDON · NEW YORK · DELHI

Anthem Press
An imprint of Wimbledon Publishing Company
www.anthempress.com

This edition first published in UK and USA 2014
by ANTHEM PRESS
75–76 Blackfriars Road, London SE1 8HA, UK
or PO Box 9779, London SW19 7ZG, UK
and
244 Madison Ave #116, New York, NY 10016, USA

First published in hardback by Anthem Press in 2013

© 2014 Valerie Purton editorial matter and selection;
individual chapters © individual contributors.

The moral right of the authors has been asserted.

All rights reserved. Without limiting the rights under copyright reserved above,
no part of this publication may be reproduced, stored or introduced into
a retrieval system, or transmitted, in any form or by any means
(electronic, mechanical, photocopying, recording or otherwise),
without the prior written permission of both the copyright
owner and the above publisher of this book.

British Library Cataloguing-in-Publication Data
A catalogue record for this book is available from the British Library.

Library of Congress Cataloging-in-Publication Data
The Library of Congress has catalogued the hardcover edition as follows:
Darwin, Tennyson and their readers : explorations in Victorian
literature and science / edited by Valerie Purton.
  pages cm
Includes bibliographical references.
ISBN 978-0-85728-076-3 (hardcover : alk. paper)
1. English literature–19th century–History and criticism. 2.
Literature and science–Great Britain–History–19th century. 3.
Darwin, Charles, 1809–1882. 4. Tennyson, Alfred Tennyson, Baron,
1809–1892. 5. Huxley, Aldous, 1894–1963. 6. Wilde, Oscar, 1854–1900.
  I. Purton, Valerie, editor of compilation.
  PR468.S34D37 2013
  820.9'356–dc23
                    2013029740

ISBN-13: 978 1 78308 348 0 (Pbk)
ISBN-10: 1 78308 348 4 (Pbk)

Cover image by Benjamin Waterhouse Hawkins, from a plate published
in *Johnson's Natural History*, 1871.

This title is also available as an ebook.

CONTENTS

Introduction   vii
Valerie Purton

Chapter 1   Tennyson's 'Locksley Hall': Progress and Destitution   1
Roger Ebbatson

Chapter 2   'Tennyson's Drift': Evolution in *The Princess*   13
Rebecca Stott

Chapter 3   History, Materiality and Type in Tennyson's *In Memoriam*   35
Matthew Rowlinson

Chapter 4   Darwin, Tennyson and the Writing of 'The Holy Grail'   49
Valerie Purton

Chapter 5   'An Undue Simplification': Tennyson's Evolutionary Afterlife   65
Michiel Nys

Chapter 6   'Like a Megatherium Smoking a Cigar': Darwin's *Beagle* Fossils in Nineteenth-Century Popular Culture   81
Gowan Dawson

Chapter 7   'No Such Thing as a Flower [...] No Such Thing as a Man': John Ruskin's Response to Darwin   97
Clive Wilmer

Chapter 8   Darwin and the Art of Paradox   109
George Levine

Chapter 9   Systems and Extravagance: Darwin, Meredith, Tennyson   135
Gillian Beer

Chapter 10   T. H. Huxley, Science and Cultural Agency   153
Jeff Wallace

Notes on Contributors   167

INTRODUCTION

Valerie Purton

'I have sometimes found in a song of Tennyson the most fitting garment of a thought engendered by a generalisation of Science.'
—Richard Owen, 1859[1]

'[T]here is a community establishing itself between literature and science, and I rejoice in that community [...] for the highest aim of science and literature, is the same; it is to diffuse, to reveal and to embody truth.'
—Thomas Henry Huxley, 1860[2]

'Presented rightly to the mind, the discoveries and generalisations of modern science constitute a poem more sublime than has ever yet addressed the human imagination. The natural philosopher today may dwell amid conceptions which beggar those of Milton.'
—John Tyndall, 1863[3]

Charles Darwin and Alfred, Lord Tennyson were exact contemporaries, born in 1809, who came to have emblematic roles as representatives, respectively, of science and literature in the Victorian age. Their juxtaposition in this volume of essays is indicative of the easy commerce between literature and science during that period and provides a salutary reminder that the two categories need to be understood within their historical context rather than assumed to be trans-historical absolutes. Readers of Darwin and Tennyson included all the significant thinkers of the day, in every field. Two – John Ruskin and Thomas Henry Huxley – are given special attention in this collection, in which a range of twenty-first-century critics from various literary disciplines address issues raised by the interaction of Victorian literature and science.

A brief overview of the historical context suggests that the interpenetration of literature and science in the Victorian period was everywhere observable. Men of science were fascinated by literature; literary authors were equally drawn to science. At the beginning of Victoria's reign, science was dominated by the 'gentlemen of science', usually Oxbridge-educated members of the Church of England – men such as Charles Babbage, John Herschel, William Whewell and William Buckland. These men were not in rebellion against William Paley's natural theology, which saw the natural world as full of evidence of God's grand design. Their foundation in 1831 of the British Association for the Advancement of Science, which first gave what might be called a 'public image' to science, was in no way intended as a revolutionary act. Between 1830 and 1833, however, Buckland's student Charles Lyell published his three-volume *Principles of Geology* (in which, as a well-educated nineteenth-century intellectual, he felt it perfectly appropriate to quote liberally from Byron), and in so doing he gave impetus to ideas which were to revolutionize the imaginations of both Darwin and Tennyson. Darwin took Lyell's first volume with him when he left England on the *Beagle* in 1831; he had the second sent out to him at Montevideo in 1832; and the third he collected in Valparaiso in July 1834. (He also took with him Milton's *Paradise Lost*.) Tennyson had certainly read Lyell by 1836 when, in a letter to Richard Monckton Milnes, he paraphrased a section from book II, chapter 18.[4] In 1844 Darwin and Tennyson both read Robert Chambers's *Vestiges of the Natural History of Creation*, Darwin with disdain, Tennyson initially with a great deal of enthusiasm: he sent his publisher Edward Moxon out to buy a copy as soon as it appeared, declaring, 'it seems to contain many speculations with which I have been familiar for years, and on which I have written more than one poem.'[5] It is these speculations in Tennyson's poetry that the first four essays of the present volume examine. In 1859 the first edition of *On the Origin of Species* lay in a bookshop window alongside the first edition of the first batch of *Idylls of the King* ('Guinevere', 'Elaine', 'Vivien', and 'Edith'). The later *Idylls* are shot through with evolutionary ideas: like the post-Darwinian novels, they too provide evidence of 'Darwin's plots'.

Definitions of 'literature' and 'science' in the discourse of Victorian Britain, as the foregoing would imply, were notoriously fluid, and there was little agreement about their usage. To the Royal Literary Fund in the mid-century, 'science' was still a branch of literature – since 'literature' retained its generous eighteenth-century usage, in which it included virtually all forms of writing. When Charles Darwin in the *Origin* envisions evolutionary change, he does so in explicitly literary terms:

> I look at the natural geological record, as a history of the world imperfectly kept and written in a changing dialect; of this history we possess the last volume

alone, relating to one, two or three countries. Of this volume, only here and there a short chapter has been preserved: and of each page, only here and there a few lines. Each word of the slowly changing language, in which the history is supposed to have been written, being more or less different in the interrupted succession of chapters, may represent the apparently abruptly changed forms of life, entombed in our consecutive, but widely separated formations.[6]

The word 'scientist' itself, in something approaching its twenty-first-century sense, was only coined in the 1830s by William Whewell, Alfred Tennyson's tutor at Trinity College, Cambridge. Much scientific writing, notably that of John Tyndall, was assumed to possess an imaginative dimension and was subsumed into mid-Victorian literary culture. Intellectuals such as George Henry Lewes maintained the tradition of the Romantic poets, especially Shelley, in assuming it was possible to preserve a many-sidedness: Lewes wrote novels, plays and literary reviews, but he also conducted scientific experiments exploring the physiological basis of the mind, and published five volumes of *Problems of Life and Mind* (1874–79). His four reviews of 'Mr Darwin's Hypothesis' (1868) had, after all, been praised by Darwin himself, and it was Darwin who had encouraged him to work them into a book.[7]

Darwin and Tennyson had both encountered William Paley's *Natural Theology* (1802) as students at Cambridge. Later in their careers, they were to be painfully caught up in the eventual and inevitable rupture between science and literature. Tennyson's agonized 'evolutionary stanzas' in *In Memoriam* 54–6 and Darwin's uneasy inclusion of the phrase 'by the Creator' in the famous last sentence of the *second* edition of the *Origin*, are merely the two best-known of many examples of the authors' involvement. The 'evolutionary naturalists' who formed the second generation of scientific practitioners no longer imagined the natural world as being contained within a religious framework. Men such as Thomas Huxley, Herbert Spencer, Francis Galton and George Henry Lewes, as well as Darwin himself, aimed to build a professional discipline of science that was essentially secular in its underpinning. At the same time, they went on drawing on what, in Matthew Arnold's terms, were the moral and spiritual resources of literature to communicate their discoveries.[8]

On the other side, contemporary scientists, particularly Huxley, quickly recognized Tennyson for his ability to synthesise the new ideas of science into lines of poetry which could be understood by a worldwide readership. The lifelong friendship between Tennyson and Huxley is particularly instructive. The two men came to know each other in London in the 1860s, where they were part of a circle including Tyndall, Herschel and Norman Lockyer. Nominated by Huxley in 1864 for a fellowship of the Royal Society,

Tennyson declined, but when the invitation was repeated the following year, he accepted and was introduced to the society on 7 December 1865. Though he rarely attended subsequent meetings, his membership remained culturally significant. Edmund Lushington wrote to Emily Tennyson on 6 April 1866, quoting a recent conversation with Thomas Huxley: Huxley had talked of his 'unbounded admiration' for Tennyson and commented that, 'We scientific men claim him as having quite the mind of a man of science.'[9] In his turn, when he wrote about David Hume in the *English Men of Letters* series (1879), Huxley was described by the *Pall Mall Gazette* (1886) as being 'hardly less distinguished for culture than for science'.[10] At this point, significantly, it is 'culture' rather than 'literature' which is being constructed as 'not science'. Huxley's public reputation was greater, apparently, than the complementary role implied in the appellation 'Darwin's bulldog'. In conversation with James Addington Symonds in 1865, it was to Huxley rather than to Darwin, that Tennyson attributed the notion of man's descent from apes: 'Huxley says we may have come from monkeys. That makes no difference to me. If it is God's way of creation, He sees the whole, past, present and future, as one.'[11] There is no record of Tennyson's response in 1870 to Darwin's *Descent of Man*, although Tennyson's is the only contemporary poetry Darwin quotes in the book. On 17 March 1873, both Huxley and Tyndall called on Tennyson at Farringford on the Isle of Wight. Emily Tennyson's journal comments that 'Mr Huxley seemed to be universal in his interest and to have a keen enjoyment of life. He spoke of *In Memoriam*.'[12] By the 1880s, Matthew Arnold's attacks on Huxley over what should be included in a liberal education were read as evidence of the beginning of a complete rupture between science and literature – a rupture which culminated in the familiar 'two cultures' formulation of C. P. Snow in the 1960s. It is important to note, however, that Huxley was not himself advocating a move away from literature towards science, but rather a move from the classics to modernity: it was both modern literature and science that he proposed to add to the educational curriculum, at the expense of what he took to be too exclusive a focus on classical languages and literature.

Huxley's well-known tribute to Tennyson (discussed by Rebecca Stott in Chapter 2) suggested his optimism about a future community of literature and science: Tennyson was, he said, 'the *first* poet since Lucretius who has understood the drift of science'.[13] Immediately after Tennyson's death in 1892, Huxley wrote a subtly different and much more pessimistic version of the tribute: 'He was the *only* modern poet, in fact the only poet since the time of Lucretius, who has *taken the trouble to understand* the work and tendency of the men of science'.[14] Huxley also crafted his own four-stanza sub-Tennysonian poem, beginning, 'Bring me my dead!', including lines redolent of its subject such as 'With thoughts that cannot die' and 'Into the storied hall, / Where I have garnered all', and ending, 'the shadows closer creep / And whisper softly: All must fall asleep.'[15] His 1893 Romanes lecture builds its exordium on a rather hectic series of borrowings from *In Memoriam* and 'Ulysses':

> We have long since emerged from the heroic childhood of our race, where good and evil could be met with the same 'frolic welcome', the attempts to escape from evil, whether Indian or Greek, have ended in flight from the battlefield; it remains for us to throw aside the youthful overconfidence and the no less youthful discouragement of nonage. We are grown men, and must play the man
> > strong in will
> > To strive, to seek, to find and not to yield,
>
> cherishing the good that falls in our way, and bearing the evil, in and around us, with stout hearts set on diminishing it. So far, we may all strive with one faith to one hope:
> > It may be that the gulfs will wash us down.
> > It may be we shall touch the Happy Isles,
> > [...] but something ere the end,
> > Some work of noble note may yet be done.[16]

Darwin's reading of Tennyson seems to have been less enthusiastic and less thorough than Huxley's – although Tennyson is, as already mentioned, the only nineteenth-century poet he quotes in *The Descent of Man*. The quotation he uses is from 'Guinevere' – that early *Idyll* which was published in the same year as *On the Origin of Species*, in 1859. Thus the *Idylls* can be seen as a cultural meeting place, in which the two great Victorians, over several decades, debated and shared ideas. Darwin uses the 'Guinevere' quotation as an illustration of 'the highest stage in moral culture at which we can arrive'.[17] Tennyson, at this early stage in the *Idylls*, was actually working with the notion of progressive evolution he had found in Jean-Baptiste Lamarck and, in 1844, in Chambers's *Vestiges* (not at all highly-regarded by Darwin). Tennyson had adumbrated this same theory in a verse in *In Memoriam*, probably written in the late 1840s. This verse also proved particularly resonant for Darwin as he sought, thirty years later, a way of communicating what was in reality the much bleaker assumption underpinning the principle of natural selection. Tennyson, absorbing Chambers, adjures humanity to

> Arise and fly
> The reeling Faun, the sensual feast;
> Move upward, working out the beast,
> And let the ape and tiger die.[18]

Darwin sees an example of this 'working out the beast' in Guinevere's brave acceptance of the need to sacrifice her love of Lancelot – to 'control her thoughts' as only an advanced human being could:

> ' [...] Not ev'n in inmost thoughts to think again
> The sins that made the past so pleasant to us [...] '[19]

Gowan Dawson has argued that it may well have been Darwin's poetry-loving wife, Emma, who recommended this quotation.[20] Certainly what Darwin doesn't pick up is the characteristically Tennysonian ambiguity of the immediately succeeding lines, in which the sensuous presence of Lancelot returns, having escaped from that moralizing negative. The larger context indeed includes an earlier line which echoes King Claudius's vain attempt at repentance in *Hamlet*. The final impression is not of an advanced human being but of a desperate soul striving, almost certainly in vain:

> ' [...] But help me, heaven, for surely I repent.
> For what is true repentance but in thought –
> Not even in inmost thought to think again
> The sins that made the past so pleasant to us:
> And I have sworn never to see him more,
> To see him more.'
>   And even in saying this,
> Her memory from old habit of the mind
> Went slipping back upon the golden days [...]
> (370–77)

As in the case of Huxley's rather impressionistic use of 'Ulysses', Victorian scientists were probably as guilty of casual reading and indeed misreading of their poetic sources as Victorian poets and novelists were guilty of superficial reading of scientific material.

This issue of 'reading and misreading' is dealt with in various ways in the chapters which follow. Underlying them all is the assumption that the 'cultural interpenetration' of Victorian literature and science was made possible because the Victorian sages, as well as the wider intellectual public, were all intently, decade by decade, *reading each other*. James A. Secord's seminal *Victorian Sensation* examines the dialogic acts of reading and writing which made up mid-Victorian culture by focusing on the public reception of a single work, Chambers's *Vestiges*. Secord examines Darwin's ways of reading and broadens the argument to suggest how other scientists might also have read the poets: 'books were not for ostentatious display, but tools for use [...] Everything was aimed towards maximum efficiency in constructing and elaborating his theories'. On the other hand, 'some books were read for extraction, others for relaxation or amusement.'[21] (The notion of 'acts of reading' takes us back to Tennyson and Darwin, and to their separate readings of Charles Lyell, discussed above.) Secord focuses on diaries, letters, press reports and so on, to offer a new approach both to the history of science and to the history of reading. David Amigoni, in *Colonies, Cults and Evolution*, extends that approach to locate within the writings of a range of Victorians 'the marginal notes and asides that link them, intertextually and dialogically, into the wider *making of a culture*.'[22]

The essays that follow examine various examples of that 'making of a culture' as scholars of Darwin, Tennyson, Ruskin, Huxley, Meredith and other Victorian figures explore the easy commerce between literature and science which predated the 'two cultures'. Huxley's confident anticipation in 1860 of a community of literature and science, viewed from a century later in the 1960s, must have seemed absurd. This was the era of C. P. Snow and F. R. Leavis – protagonists in a debate which they also in many ways embodied. The 1980s, however, saw the rise of the flourishing academic subgenre of Victorian literature and science, which has given renewed currency to Huxley's notion. From Gillian Beer's *Darwin's Plots* to George Levine's *Darwin the Writer*, literary figures have been increasingly reread through their responses to scientific thinking. Tennyson's scientific interests have been thoroughly examined, and Charles Darwin himself has been reread not only as a scientist, but as a reader of literature and a literary stylist.

A necessarily brief sketch of the development of the field of Victorian literature and science begins with Tess Cosslett's *The Scientific Movement and Literature* (1982). Gillian Beer's *Darwin's Plots* (1983; 3rd edition, 2009) considered the responses of Victorian writers including George Eliot, Charles Kingsley and Thomas Hardy to Darwinian ideas, while also considering more broadly the manifestations of evolutionary thinking in the culture of the time. George Levine developed these insights further in *Darwin and the Novelists* (1988), and in *Darwin Loves You* (2009) and *Darwin the Writer* (2011) he offered the notion of a 'two-way traffic', looking at Darwin's reading of contemporary literature and his struggle to use the language of his time in his scientific thinking. James Secord's *Victorian Sensation* (2000) examined minutely the way in which one particular Victorian work, Robert Chambers's *Vestiges*, was actually read and absorbed by readers across Victorian culture. This renewed stress on readers and on acts of reading lies behind the choice of title for the present volume. The new subgenre developed in a variety of ways: by the mid-1980s there was a move towards reading literature and science as parallel discourses; in the 1990s, there was an expansion of interest, beyond evolutionary biology and towards mind sciences; while in the early years of the twenty-first century there has been an explosion of interest in the methods by which nineteenth-century scientific ideas were transmitted. A very limited selection of significant works illustrates these broad trends: Gillian Beer, *Open Fields: Science in Cultural Encounter* (1999); Helen Small and Trudi Tate, eds, *Literature, Science, Psychoanalysis: Essays in Honour of Gillian Beer* (2003); Rebecca Stott, *Darwin and the Barnacle* (2004); Geoffrey Cantor et al., *Science in the Nineteenth-Century Periodical* (2004); Jonathan Smith, *Charles Darwin and Victorian Visual Culture* (2006); Gowan Dawson, *Darwin, Literature and Victorian Respectability* (2007); David Amigoni, *Colonies, Cults and Evolution* (2007); Ralph O'Connor, *The Earth on Show* (2007); Bernard Lightman, *Victorian Popularisers of Science* (2007); John Holmes, *Darwin's Bards: British and American Poetry in the Age of Evolution* (2009); Charlotte Sleigh, *Literature and Science* (2010); Bruce Clarke and Manuela Rossini, *The Routledge Companion to Literature and Science* (2012); and Sally Shuttleworth, *Culture and Science in the Nineteenth-Century Media* (2004) and *The Mind of the Child: Child Development in Literature, Science and Medicine, 1840–1900* (2010).

This collection begins with four essays which examine Tennyson's engagement with scientific debates and with scientists, progressing chronologically from 'Locksley Hall' (1832) through *The Princess* (1846) and *In Memoriam* (1850) to 'The Holy Grail' (1867). The pivotal fifth chapter looks at the opposite direction of the 'two-way traffic', examining how scientists read Tennyson. The second section of the book consists of four essays on Darwin, while in the final chapter, Jeff Wallace gives a fresh perspective on the 'Victorian literature and science' debate with a warning to twenty-first-century scholars against reading the role of the Victorian scientist through twenty-first-century eyes; in doing so, he ends the volume where this introduction began, with Thomas Henry Huxley.

Synopses of Chapters

Chapter 1: Roger Ebbatson – Tennyson's 'Locksley Hall': Progress and Destitution

Tennyson's 'Locksley Hall' (published 1842) was composed in the late 1830s, at a time of unprecedented social upheaval. The poem precariously balances utopian and quasi-evolutionary visions of the future against an ominous sense of crisis. Tennyson's protagonist seeks a palliative for the evils of mid-Victorian materialism by espousing a doctrine of progressive evolution and communal purpose akin to the thrust of contemporary 'scientific' texts such as Chambers's *Vestiges of Creation*. This 'upward' trajectory is undermined by the poem's conclusion, which, with its sense of millenarian ruination, speaks to Walter Benjamin's thesis that 'the concept of progress must be grounded in the idea of catastrophe.' The sense of evolutionary reversion, or Spencerian 'degeneration', is further elaborated in 'Locksley Hall Sixty Years After', in which the cry of 'Forward! Forward!' is lost within the growing gloom. Both poems thus debate the notion of 'Evolution ever climbing after some ideal good, / And Reversion ever dragging Evolution in the mud.'

Chapter 2: Rebecca Stott – 'Tennyson's Drift': Evolution in The Princess

Huxley's compliment to Tennyson, that he was 'the first poet since Lucretius who has understood the drift of science', includes a very Tennysonian word, 'drift'. In *The Princess: A Medley* (1847), Tennyson's experiment in dialogic or conversational form seeks to show that it is educated *mixed-sex* conversation that determines and shapes the drift of science. Earlier, at Cambridge, Tennyson had encountered 'Transformism', via Tiedemann and Lamarck. The young Darwin, meanwhile, was discussing Lamarck's ideas with Robert Grant at Edinburgh Medical School. Both Darwin and Tennyson went on to read Charles Lyell's *Principles of Geology* (1830–34) and then to respond in different ways to Robert Chambers's *Vestiges of the Natural History of Creation* (1844). *Vestiges* was discussed with horror and fascination at the dinner tables, mechanics institutes and salons of Britain and Europe for a considerable time. The anonymous author proposed that the earth had started out as a nebular fire mist and that all life forms on the planet had evolved from earlier simpler forms, many of them aquatic. Darwin reacted to the opprobrium meted out to the anonymous author by returning to the small-scale, to the barnacle; Tennyson responded, in *The Princess*, by embracing new ideas and new forms. The chapter proposes that the conversational form that drives the poem and its politics (between the present-day undergraduates and between the prince and the princess) was shaped by Tennyson's observations of the kinds of passionate conversations opening up around him about the new science.

Chapter 3: Matthew Rowlinson – History, Materiality and Type in Tennyson's In Memoriam

English lexicography struggles with the noun 'type'. The *Oxford English Dictionary* cites John Stuart Mill's *Logic* (1841) as the first instance of what has become the dominant sense as 'the general form, structure, or character distinguishing a particular [...] class of beings or objects.' Mill was using a definition propounded by William Whewell in his *Philosophy of the Inductive Sciences* (1840). Another sense of the term (which the *OED* attests as only twentieth-century), as signifying 'the sort of person to whom one is attracted', is in fact used by George Eliot in *The Mill on the Floss* (1860), though only as a conscious Gallicism. Whewell's use too depends upon a French source in the taxonomic theory of Georges Cuvier in *Le règne animal* (1817) and *Histoire naturelle des poissons* (1828–33). It seems likely that both Cuvier and Whewell are influenced by the fact that since the Renaissance ,'type' has been the normal French translation of the Greek 'eidea', usually rendered in English as 'form'. The Greek word is derived from the verb 'to strike' and refers to the raised image on a coin, produced by striking with a hammer (a sense preserved in modern English in 'printers' type'). Tennyson's masterpiece *In Memoriam* is widely accepted as being a typological poem. This chapter re-examines Tennyson's references to the type-concept to show that, when read in the conflicting numismatic, taxonomic and erotic contexts that the term brings with it, they are far more heterogeneous and unsettling than has yet been seen.

Chapter 4: Valerie Purton – Darwin, Tennyson and the Writing of 'The Holy Grail'

Earlier chapters have explored Tennyson's response to pre-Darwinian evolutionary debates. Chapter 4 focuses instead on the only documented meeting between the two men, on 17 August 1868, and Tennyson's subsequent completion of the *Holy Grail* Idyll in what Emily Tennyson described as 'a breath of inspiration' during the following three weeks. The Holy Grail episode had been on Tennyson's mind for over a decade: to him it was the key to the whole *Idylls* cycle, but he demurred year after year, doubting, he said, 'whether such a subject could be handled these days without a charge of irreverence.' Almost in the same breath, however, he argued that his problem was quite the opposite: that in Malory's time the task was easier because 'in those days people actually *believed* in the Grail.' The chapter explores the state of the religion and science debate in the 1860s to explain Tennyson's difficulties. It then turns to Tennyson's reading of *On the Origin of Species* in November 1859 'with intense interest' and rereads 'The Holy Grail' through the prism of the *Origin*, concluding that, at the very least, Darwin may have contributed to Tennyson's more sceptical reading of the Grail legend.

Chapter 5: Michiel Nys – 'An Undue Simplification': Tennyson's Evolutionary Afterlife

When interpreting the ethical implications of evolutionary theory in the late nineteenth century, Thomas Huxley invoked Tennyson. His moralistic rhetoric drew heavily on martial tropes and antagonistic modes of experience, with Tennyson's verse serving as a major source of inspiration. Huxley's own poetical tribute to Tennyson, composed immediately after the Poet Laureate's funeral in October 1892, characteristically praised responsible citizenship and heroic defiance of the individual's inevitable fate. However, there is an ambivalence in Huxley's approach: in 'Evolution and Ethics'(1893), he observes of ethical society that, ideally, it 'repudiates the gladiatorial theory of existence', while in the same breath he affirms that 'once for all, the ethical progress of society depends, not on imitating the cosmic process, still less in running away from it, but in combating it.' This chapter analyses 'Evolution and Ethics', examining in particular the literary texts invoked by Huxley: these include a variety of self-reflexive quest narratives ranging from Seneca, the myth of Sisyphus and the folktale of Jack and the Beanstalk, through the Book of Job and the tragedies of Oedipus and Hamlet to, finally, Robert Browning's 'Childe Roland' and Alfred Tennyson's 'Ulysses', both of which sought to challenge the mid-Victorian reader by means of a complex treatment of 'the gladiatorial theory of existence.'

Chapter 6: Gowan Dawson – 'Like a Megatherium Smoking a Cigar': Darwin's Beagle Fossils in Nineteenth-Century Popular Culture

In 1836 Charles Darwin, recently returned from his *Beagle* voyage, presented the fragmentary remains of the sloth-like creature that Cuvier had named the megatherium to the Hunterian Museum at the Royal College of Surgeons, where they were examined by Richard Owen, the Museum's rising star of comparative anatomy. Owen used them, famously, to vindicate the power of inductive reasoning by arguing for a new functional interpretation of the relation between the megatherium's anatomy and its peculiar feeding habits. From the 1840s onward the megatherium became a celebrated figure in nineteenth-century culture. In an era of enormous social, technological and cultural change, the lumbering but seemingly perfectly adapted creature, reconstructed from tiny fragmentary parts, offered ways of understanding novel technologies such as railway locomotives – described in *Fraser's Magazine* as resembling 'a megatherium smoking a cigar' – or new publishing forms such as the lengthy novels read in small serial parts that were frequently described as types of megatherium by both critics and novelists. Ranging from Charles Kingsley's *Alton Locke* to William Makepeace Thackeray's *The Newcomes*, to Benjamin Waterhouse Hawkins's prehistoric models at the Crystal Palace and Victorian concerns about slothfulness, this chapter examines how Darwin's fossil samples from the *Beagle* took on a life of their own in nineteenth-century culture.

Chapter 7: Clive Wilmer – 'No Such Thing as a Flower [...] No Such Thing as a Man': John Ruskin's Response to Darwin

Whatever the differences may have been between Darwin's theory of natural selection and the notion of evolution that the undergraduate Tennyson supported in a debate, it is clear that the poet was well-prepared for *On the Origin of Species*. The young John Ruskin, by contrast, brought up as a strict Evangelical and taught at Oxford by William Buckland of *The Bridgewater Treatises*, was committed to Natural Theology from the outset. From the first appearance of the *Origin*, he abused Darwinism and, in particular, theories of competition whenever opportunity occurred. There is, however, another side to the story: Ruskin from his teens was an enthusiastic student of geology and certainly understood the implications of Lyell's *Principles of Geology*. He was also familiar with a range of modern scientific thought, from Cuvier to Louis Agassiz, the atmosphere in which Darwin's theory was born. His attitude to nature – a close attention to its particulars and a realist understanding of natural forms – belongs to much the same tradition as Darwin's. It is no accident that when the two men met, they found they shared many enthusiasms. This chapter argues that Ruskin's response to Darwinism was less an intellectual disagreement with the theory than an impassioned reaction to what he saw, with visionary intensity, as its implications for the future of humanity.

Chapter 8: George Levine – Darwin and the Art of Paradox

The chapter addresses not so much evolutionary ideas as the form and language of Darwin's writing, examining its influence on an unexpected range of writers, from Arthur Conan Doyle to Walter Pater and Oscar Wilde. Fully to grasp the art of Darwin's prose requires a very modernist shift in point of view. His new sublime is not so much outside, in the wonders of nature he so much admired and felt, as inside, in the power of mind to imagine beyond what it sees. In rejecting the traditional anthropocentric view of the universe, Darwin had to struggle with a language that seemed to reflect nature as it was; in doing so he developed a prose that often took the form of paradox. 'Natural history,' said G. H. Lewes in 1860, 'is full of paradoxes'.

This chapter looks back to Darwin's counterintuitive vision and forward to the more demonstratively paradoxical modes of the fin de siècle. Darwin's vision of the world is seen not as tragic but as comic in its radical reversal of our sense of things. The best locus for articulating the aesthetic of post-Darwinian literature and its inward and paradoxical turn is in Oscar Wilde's 'Decay of Lying', in which Vivian builds his theory of art out of Darwinian materials, laughing brilliantly along the way.

Chapter 9: Gillian Beer – Systems and Extravagance: Darwin, Meredith, Tennyson

Just as the *sublime* is key to Romantic sensibility, *extravagance* is its transformed equivalent in the later nineteenth century. Darwin's thinking is itself extravagant and is based on the principle of extravagance, of an excess number of individuals being produced to help a species survive. Where to Malthus such proliferation was a waste of energy, to Darwin it is a delight. He sees the 'endless forms of the world as "most beautiful"': his is a system which demands extravagance. To Tennyson the extinction of species seemed more heartbreaking than it did to Darwin. In a late poem, 'The Islet', he grasps Darwin's sense of the meagreness of isolation – the horror of one bird, one single note, one serpent. Darwin's poets, however, were the earlier generation – Byron, Wordsworth, Thomson, Shelley and Keats. Later, his enthusiasm for poetry vanished, though he is bound to have been aware of *In Memoriam* (1850); and in the death of his daughter Annie in 1851, he, like Tennyson knew the extravagance of loss. Tennyson encouraged George Meredith in his early poetry, and it was Meredith whose poetic career took in the full impact of Darwin's ideas. In *The Egoist* (1879), Meredith plays wild games with Darwin's arguments in *The Descent of Man* (1871). Darwin's comments on the intricacies and extravagance of birdsong are embodied in Meredith's 'The Lark Ascending', as well as in Vaughan Williams's later musical setting. Extravagance is, for Darwin, Tennyson and Meredith, a way of imagining the world at full stretch and watching it change.

Chapter 10: Jeff Wallace – T. H. Huxley, Science and Cultural Agency

In a pioneering study of T. H. Huxley published in 1978, James Paradis makes the claim that Huxley, in his writing and public speaking, created a 'unique cultural agent' – 'the scientist'. This chapter explores and questions the concept of cultural agency as it bears on recent critical debate around Huxley's life and work. Focussing on the practice of what Adrian Desmond calls 'the new contextual history of science', the chapter examines the implied relationship, in the arguments of scholars such as Paradis and Desmond, between textuality and rhetorical skill in science on the one hand, and scientific epistemology on the other. How far might the concept of cultural agency encourage precisely the kind of reductive polarization of science that the new contextual histories should actually discourage? How easy is it to assume that culture is a richer domain than science? Exemplifying the lure of this assumption in one of Wallace's own critical exchanges, the chapter uses the arguments of the critic Neil Belton as a counterweight. In the second half of the chapter, an analysis of a range of Huxley's writings from the 1860s through to the 1890s tests out Belton's idea of Huxley's 'creative rationalism', within which science might be seen as the driving force of cultural agency, before it is seen as the literary or textual product of cultural agency.

***

One issue much discussed in recent contributions to the 'Victorian literature and science' debate is the ordering of nouns within the phrase itself: George Levine raised the question in *One Culture: Essays in Science and Literature* (1987): '"And" cloaks many different sorts of relationships. If we think of "influence" in this connection, we normally think of science influencing literature [...] But the influence works the other way too, as strong developments in externalist history of science have been demonstrating.'[23] The debate has come a long way since then. Ralph O'Connor was of a faction who believed it might be better for Victorianists to think of 'science *as* literature, rather than science *and* literature'.[24] Dawson and Levine returned to the question in 2007–2008 in the *Journal of Victorian Culture* (vols11.2 and 12.1 respectively): Levine summarises Dawson's argument, as questioning 'the tendency of interdisciplinary scholars, eager to find connections among activities that had traditionally been thought to be heterogeneous, to assume a "common context" for science and literature, and to overlook the ways in which science and literature were, in certain respects, after all antagonistic.' Dawson and Lightman, in the introduction to their *Victorian Science and Literature* volume for Pickering and Chatto (2012), register their awareness of the debate about appropriate word order by suggesting that 'some readers may wish to place their imaginary quotation marks' around the anthology's 'avowedly problematic title'.[25] Lightman goes on to suggest that his own *Victorian Popularisers of Science* (2007) attempts to 'dispense with any lingering disciplinary distinctions and [to] speak equally to historians of science as much as to literary critics.'[26] In eventually making the editorial decision simply to follow the alphabet for the present volume, both in the surnames (in the title) and in the subject names (in the subtitle),

I have the advantage of achieving a degree of even-handedness, since this method gives precedence to the scientist in the title and to literature in the subtitle. In staying with the 'literature and science' formulation, though, the collection must begin with essays on Tennyson.

Notes

1. Richard Owen, Royal Literary Fund Annual Report 10 (1859), 27 RLF Archive, Loan 96, British Library. I would like to thank Dr Gowan Dawson for drawing my attention to this reference in his paper given at the Birkbeck Dickens Day, 10 October 2009.
2. Royal Literary Fund Annual Report 10 (1860), 32.
3. John Tyndall, *Heat Considered as a Mode of Motion* (1863), 433, quoted in Robin Gilmour *The Victorian Period: The Intellectual and Cultural Context of English Literature 1830–1890* (London: Longman, 1993), 142.
4. Alfred Tennyson, *The Letters of Alfred, Lord Tennyson*, ed. Cecil Y. Lang and Edgar F. Shannon, Jr, 3 vols (Oxford: Oxford University Press, 1981–90), 1:145.
5. Tennyson, *Letters*, 1:230.
6. Charles Darwin, *On the Origin of Species*, ed. Gillian Beer (Oxford: Oxford University Press, 2008), 229.
7. See Rosemary Ashton, *George Henry Lewes: A Life* (Oxford 1991), 244–5, cited in Gilmore 143.
8. See for example the preface to the first edition of *Poems* (1853) (Harmondsworth: Penguin, 1970), 53.
9. Edmund Lushington to Emily Tennyson, 6.4.66, Emily Tennyson's Journal, Tennyson Research Centre.
10. *Pall Mall Gazette*, 2 October 1886, cited in David Amigoni, *Colonies, Cults and Evolution* (Cambridge: Cambridge University Press, 2007), 175.
11. Tennyson, *Letters*, 2:418.
12. Hallam Tennyson, *Alfred, Lord Tennyson: A Memoir by His Son*, 2 vols (London: Macmillan, 1897), 2:143.
13. Thomas Henry Huxley, *Life and Letters of Thomas Henry Huxley*, ed. Leonard Huxley, 2 vols (London: Macmillan, 1900), 2:338 (my italics).
14. Ibid., 2:337 (my italics).
15. The poem, entitled 'Westminster Abbey: October 12 1892', appeared as the first item in a collection entitled: 'To Tennyson: The Tributes of His Friends', in *Nineteenth Century* 32, July–December 1892 (London: Sampson, Low, Marston and Company, 1892), 831–2.
16. *Evolution and Ethics and Other Essays* (Teddington, Middlesex: Echo Library, 2006), 44–5.
17. Charles Darwin, *The Descent of Man and Selection in Relation to Sex* (London: John Murray, 1871), 1(3):101.
18. Alfred Tennyson, *Idylls of the King* (London: Edward Moxon, 1859), 250.
19. Ibid., 244–5.
20. Gowan Dawson, *Darwin, Literature and Victorian Respectability* (Cambridge: Cambridge University Press, 2007), 53.
21. James A. Secord, *Victorian Sensation: The Extraordinary Publication, Reception, and Secret Authorship of 'Vestiges of the Natural History of Creation'* (Chicago: University of Chicago Press, 2000), 427–8.

22 David Amigoni, *Colonies, Cults and Evolution* (Cambridge: Cambridge University Press, 2007), 9 (my italics).
23 George Levine, *One Culture: Essays in Science and Literature* (Madison: University of Wisconsin Press, 1987), 6.
24 Ralph O'Connor, *The Earth on Show: Fossils and the Poetics of Popular Science, 1802–1856* (Chicago and London: University of Chicago Press, 2007), 15.
25 Gowan Dawson and Bernard Lightman, eds, *Victorian Science and Literature*, 8 vols (London: Pickering and Chatto, 2012) 1:4.
26 Ibid.

Chapter 1

TENNYSON'S 'LOCKSLEY HALL': PROGRESS AND DESTITUTION*

Roger Ebbatson

The composition of Tennyson's 'Locksley Hall' during 1837–38 coincided with the foundation of the Corn Law League, the promulgation of the People's Charter and the controversy over the enforcement of the New Poor Law, whilst its publication in 1842 was marked by the riots over the rejection of the Chartist petition. These five years have been characterized as 'the grimmest period in the history of the nineteenth century', a moment when 'Industry came to a standstill, unemployment reached hitherto unknown proportions, and with high food prices and inadequate relief the manufacturing population faced hunger and destitution.'[1] Tennyson's poem is precariously balanced between utopian and scientifically orientated visions of the future – as when the feverish protagonist recounts how he dipped 'into the future far as human eye could see' and 'Saw the vision of the world and all the wonder that could be' (15–16)[2] – and an ominous sense of social change: 'Slowly comes a hungry people, as a lion creeping nigher, / Glares at one that nods and winks behind a slowly-dying fire' (135–6). The predominant mood is misanthropic, the hero urging his army companions to leave him alone to contemplate the 'dreary gleams about the moorland flying over Locksley Hall' (4) and complaining of the 'social wants' (59) and 'social lies' (60) of contemporary society. The poem is immersed in a twilit atmosphere in which the springtime joy of the protagonist's love for his cousin is transmuted, following his rejection and her marriage to an upper-class suitor, into its opposite:

O my cousin, shallow-hearted! O my Amy, mine no more!
O the dreary, dreary moorland! O the barren, barren shore!
(39–40)

---

* A version of this chapter first appeared as 'Tennyson's "Locksley Hall": Progress or Destitution' in Roger Ebbatson's *Landscape and Literature 1830–1914: Nature, Text, Aura* (Basingstoke: Palgrave Macmillan, 2013) reproduced with permission of Palgrave Macmillan. The full published version of this publication is available from http://www.palgraveconnect.com/pc/doifinder/10.1057/9781137330444.

The contradictory valences of this puzzling text, with its enthusiasm for the future and vituperative critique of the present, are mirrored and refracted in Martin Heidegger's seminal essay, 'What Are Poets For?', which examines the role of the poet in what Heidegger designates 'a destitute time'.[3] In the era of 'the default of God', Heidegger postulates, 'the divine radiance has become extinguished in the world's history' (*PLT*, 89), just as in 'Locksley Hall' the sight of the Pleiades 'rising through the mellow shade' (9) gives way to a vengeful 'vapour from the margin, blackening over holt and heath' (191, 193). The present age, in Heidegger's view, is poised over existential 'cliffs of fall': 'In the age of the world's night, the abyss of the world must be experienced and endured,' and for this 'it is necessary that there be those who reach into the abyss' (*PLT*, 90). Under this analysis it is the poet who is enabled to plunge creatively into the abyss: 'To be a poet in a destitute time means: to attend, singing, to the trace of the fugitive gods' (*PLT*, 90), just as Tennyson fashioned poetry for the 1840s so that, as Heidegger says of Rilke, 'Song still lingers over [the] destitute land,' and 'the song still remains which names the land over which it sings' (*PLT*, 94, 95).

In the backstory of the poem, the hero's youthful love for his cousin Amy is shattered by her marriage to the lord of the manor:

> He will hold thee, when his passion shall have spent its novel force,
> Something better than his dog, a little dearer than his horse.
> (49–50)

This imbroglio leads both to fantasies of a *liebestod*, or love-death, in which the two cousins are imagined 'Rolled in one another's arms, and silent in a last embrace' (58), and to the hero's consequent rejection of a materialist age:

> Cursèd be the social wants that sin against the strength of youth!
> Cursèd be the social lies that warp us from the living truth!
> (59–60)

From the bilious perspective of the hero, this is an era dominated by property and the marriage market:

> What is that which I should turn to, lighting upon days like these?
> Every door is barred with gold, and opens but to golden keys.
> Every gate is thronged with suitors, all the markets overflow.
> I have but an angry fancy: what is that which I should do?
> (99–102)

Catherine Hall has pertinently noted, apropos of the condition-of-England novel, how the private or domestic world of love and marriage 'is often set aside for the alleviation of antagonisms that cannot be resolved in the social world', but she adds that the realist novel seeks to 'connect public and private fields' in ways which indicate 'a deep rift between them'.[4] As Heidegger phrases it in his diagnosis of destitution, 'the humanness of man and the thingness of things dissolve into the calculated market value of a market which [...] spans the whole earth,' with the effect that all beings become subject 'to the trade of a calculation that dominates' (*PLT*, 112). Tennyson's protagonist seeks hectically for a remedy and initially discovers one in the liberal doctrine of progress and communal purpose:

> Men, my brothers, men the workers, ever reaping something new:
> That which they have done but earnest of the things that they shall do.
> (117–18)

This energized and rhythmical declaration, as Kirstie Blair observes, 'sounds not unlike the militant marches of Chartist poetics',[5] but it ushers in the well-known evocation of the emergence of a Saint-Simonian future of world trade out of a phase of aerial conflict:[6]

> For I dipt into the future, far as human eye could see,
> Saw the Vision of the world, and all the wonder that would be;
>
> Saw the heavens fill with commerce, argosies of magic sails,
> Pilots of the purple twilight, dropping down with costly bales;
>
> Heard the heavens fill with shouting, and there rained a ghastly dew
> From the nations' airy navies grappling in the central blue;
>
> Far along the world-wide whisper of the south-wind rushing warm,
> With the standards of the peoples plunging through the thunder-storm;
>
> Till the war-drum throbbed no longer, and the battle-flags were furled
> In the Parliament of man, the Federation of the world.
> (119–28)

Tennyson's poem, in this remarkable passage, thus speaks to the aspirations and tensions of its moment, in which the new theory of free trade, focused upon the anti–Corn Law campaign, replaced the dogma of national economic competitiveness and rivalry. As enunciated by Lecky and others, the economic benefits of peace were an overriding consideration; Alan Swingewood has suggested, apropos of this period,

> Social and economic theory [...] tended to eliminate contradictions in favour of evolution and progress, and beginning with the Chartist movement in the 1830s bourgeois social theory is forced to see the 'social problem' increasingly in ideological terms.[7]

Tennyson's speaker functions as a kind of Heideggerian 'precursor', one who 'arrives out of [the] future, in such a way that the future is present only in the arrival of his words' (*PLT*, 139). Through an act of 'ideological misrecognition' prompted by his ill-fated love, the hero adopts a posture whose ecstatic image of the future is dialectically posited upon, and undermined by, his inability to cope with the present:

> So I triumphed ere my passion sweeping through me left me dry,
> Left me with the palsied heart, and left me with the jaundiced eye;
>
> Eye, to which all order festers, all things here are out of joint
> (131–3)

In the abrupt mood changes, aptly mirrored in the headlong trochaic rhythmic pattern, with the vision of progress rapidly dissolving at the prospect of the Chartist insurgence – that 'hungry people, as a lion creeping nigher' (135) – we discern what might be termed the liquidation of dramatic monologue, in which the single voice splinters into a spasmodic cacophony of warring tones. Indeed, the protagonist's self-division unwittingly mimics the class tensions of the poem's historical moment: as Anne Janowitz has noted, 'by 1842, the term "the people" was primarily used by parliamentary politicians to describe the lower orders.'[8] Patrick Joyce has demonstrated the complexities of the terminology deployed in this debate, but he confirms that the identity of 'the people' 'could in fact take on a class character, turning upon the idea of labour as a "working class" in conflict with capital'. The notion of 'the people', Joyce further claims, could function as 'a principle of social exclusion as well as of social inclusion, as the "working class" increasingly stood proxy for the nation'.[9]

The oscillations in the mind of the protagonist come to a head in the subsequent desert-island fantasy. The hero's colonial origins mark him out as a figure prone to atavistic longings and lead him to call 'for some retreat / Deep in yonder shining Orient, where my life began to beat' (153–4). His father, we learn, fell 'in wild Mahratta battle cry', leaving the hero 'a trampled orphan, and a selfish uncle's ward' (156). Nineteenth-century India,

Homi Bhabha has argued, represented 'the perpetual generation of a past–present which is the disturbing, uncertain time of the colonial intervention and the ambivalent truth of its enunciation'.[10] Locksley Hall itself, as a building, thus comes to represent the haunted Otherness pertaining to colonial history, a history in which the law of the Father is constantly redefined, undermined or hybridized in a process through which the colonizer becomes, as it were, orphaned to himself. Tennyson's Oedipal variant fuels the hero's Stevensonian desire to 'burst all links of habit':

> [...] there to wander far away,
> On from island unto island at the gateways of the day.
> (157–8)

Here, under 'Breadths of tropic shade' (160), where 'never floats an European flag' (161), life appears to offer a Lotos-like refuge from modernity:

> Droops the heavy-blossomed bower, hangs the heavy-fruited tree –
> Summer isles of Eden lying in dark-purple spheres of sea.
> (163–4)

There would be more 'enjoyment' in this enervated paradise, he reflects, than in the technological 'march of mind' epitomized 'In the steamship, in the railway, in the thoughts that shake mankind' (165–6). At the heart of this dream is a powerful erotic dimension:

> There the passions cramped no longer shall have scope and breathing space;
> I will take some savage woman, she shall rear my dusky race.
> (167–8)

As Patrick Brantlinger has remarked, the exotic other, as desert island or mysterious Orient, 'seems to the early Tennyson a daydream realm of ahistorical, exotic, and erotic pleasures'.[11]

The daydream is soon shattered, however, as the hero reverts to liberal race orthodoxy, scoffing at the notion that he should

> [...] herd with narrow foreheads, vacant of our glorious gains,
> Like a beast with lower pleasures, like a beast with lower pains!
> (175–6)

In examining the contradictory implications of the doctrine of progress, T. W. Adorno contends that it is not 'man's lapse into luxuriance that is to be feared' but rather what he terms 'the savage spread of the social under the mask of universal value, the collective as a blind fury of activity'.[12] However, Robert Knox's pseudo-scientific argument in *The Races of Man*, published in 1850, that 'The Saxon will not mingle with the dark race,'[13] is endorsed by Tennyson's hero and foregrounded in his proclaimed inability to be 'Mated with a squalid savage' (177). To the contrary, as 'heir of all the ages' (178), the white European male feels bound to embrace the progressive sense of a futurity guaranteed by scientific innovation:

> [...] Forward, forward let us range,
> Let the great world spin for ever down the ringing grooves of change.
> (181–2)

Charles Kingsley averred that the final movement of 'Locksley Hall' spoke of 'man rising out of sickness into health', 'conquering his selfish sorrow' and expressing 'faith in the progress of science and civilisation, hope in the final triumph of good'.[14] But although he affirms that it is better to contemplate 'fifty years of Europe than a cycle of Cathay' (184), the speaker's investment in the doctrine of progress is verbally shadowed by intimations of calamity:

> Rift the hills, and roll the waters, flash the lightnings, weigh the Sun.
> (186)

Tennyson's vertiginous text ends, indeed, with a sense of ruination and millenarian apocalypse which is very much of its period:

> Howsoever these things be, a long farewell to Locksley Hall!
> Now for me the woods may wither, now for me the roof-tree fall.
>
> Comes a vapour from the margin, blackening over heath and holt,
> Cramming all the blast before it, in its breast a thunderbolt.
>
> Let it fall on Locksley Hall, with rain or hail, or fire or snow;
> For the mighty wind arises, roaring seaward, and I go.
> (189–94)

The psychic problems of 'Locksley Hall', though ostensibly rooted in erotic failure, may be read as an *effect* of social problems and class conflict, and the fragmentary form itself, in voicing what Strindberg would designate the 'split and vacillating' personality of modernity,[15] demonstrates how an emergent psychological domain required new lyric conventions to embody its effects.

In imagining warfare or revolutionary insurrection, as Brantlinger has suggested, 'Locksley Hall' conforms to a general pattern in which Tennyson 'juxtaposes peace and war in ways that frequently associate the former with cowardice and greed, the latter with the highest virtues'.[16] Furthermore, in veering between a sense of social cohesion and an alienated selfhood, the text mirrors what Janowitz, in her discussion of Chartist poetry, terms 'the contest of individualist and communitarian poetics'.[17]

The complex valences of Tennyson's conclusion gesture towards a dissatisfaction with the liberal political dogma attendant upon an 'age of transition', and the hero's final utterance might be weighed against Walter Benjamin's notation of the way in which 'The concept of progress must be grounded in the idea of catastrophe.'[18] According to Benjamin's diagnosis, a materialist critique 'blasts the epoch out of the reified continuity of history'. Just as the hall's roof-tree is destined to 'fall', so a materialist reading of history 'explodes the homogeneity of the epoch, interspersing it with ruins – that is, with the present' (*AP*, 474). For Tennyson's protagonist the wind 'arises, roaring seaward', and in Benjamin's account the dialectician must 'have the wind of world history in his sails' in order 'to dissipate the semblance of eternal sameness, and even of repetition, in history' (*AP*, 473). In every true work of art, Benjamin contends, 'there is a place where, for one who removes there, it blows cool like the wind of a coming dawn' (*AP*, 474). This concept is echoed or voiced through Tennyson's extraordinary verse form, its rolling trochees paradoxically bringing into being what Benjamin terms a 'caesura in the movement of thought' which embodies a 'violent expulsion from the continuum of historical progress' (*AP*, 475). Just as Tennyson's protagonist hails the 'flash' of the 'lightnings', so for Benjamin the dialectical reversal of a scientifically authorized liberal progress 'emerges suddenly, in a flash', 'an image flashing up in the now of its recognisability' (*AP*, 473). The hero's language in this peroration points ambiguously towards either international warfare or, closer to home, a proletarian uprising which will arise to devastate the nascent capitalist system. The poem thus speaks to a kind of Benjaminian 'constellation' of political and personal concerns, and in so doing it endorses Benjamin's claim that 'form in art is distinguished by the fact that it develops new forms in delineating new contents' (*AP*, 474). For Benjamin, the dialectical image is freighted with intimations of redemptive longing whilst symbolizing the failure to fulfil such hopes. Tennyson's poem stages and foreshadows the Benjaminian idea of a homogeneous empty time that is filled in by the ineluctable mid-Victorian belief in progress embodied in science, technology, evolutionism and the philosophy of history. The aesthetic implications of this structure of feeling are fruitfully developed by Adorno, whose definition of commodity culture as a delusional expression of collective fantasies sheds a critical light on Tennyson's text. For Adorno, the genuinely new artwork serves as 'an ominous warning, a script that flashes up, vanishes, and indeed cannot be read for its meaning'.[19] Art, which is 'profoundly akin to explosion', aspires 'not to duration but only to glow for an instant', and the artwork thus comprises 'a form of reaction that anticipates the apocalypse' (*AT*, 112). Tennyson's hero, and his textual embodiment as a 'printed voice', that is to say, is haunted by a Benjaminian sense of the loss of auratic resonance or value. The inaugural glow of 'great Orion sloping slowly to the West' (8) and of the Pleiades which 'Glitter like a swarm of fire-flies' (10) fades away to be transmuted into the 'lightnings', 'blast' and 'thunderbolt' of the conclusion. In his examination of Chartist poetry, Michael Sanders notes how the movement was 'frequently represented as an irresistible natural force' in a trope which may be characterized as 'the archetype of the destruction of the old corrupt order'.[20] In calling for the ruination of Locksley Hall, the protagonist seems to reject the utopian bourgeois dogma of progress, international trade and technical mastery of nature espoused earlier in the poem. Such a scenario postulates a violent remedy for the supremacy of commodity culture and exchange value from which Tennyson as Laureate would both suffer and profit. As Adorno phrases it, 'if artworks shine, the objectivation of aura is the path by which it perishes' (*AT*, 112).

The implications of 'Locksley Hall' are thus allegorical in their examination of the poet's predicament in a destitute time. In Adorno's view, 'Not only are artworks allegories, they are the catastrophic fulfilment of allegories' (*AT*, 112), and his claim that 'History is the content of artworks' (*AT*, 112) is particularly relevant to a reading of this poem, furrowed as it is not only by the biographical implications of the Rosa Baring affair and the disinheritance of the Somersby Tennysons, but also by traces of the condition-of-England debate. And yet, beyond the acknowledgement of its historical context, 'Locksley Hall' offers a reading experience of baffling undecidability which arises out of a barely articulated clash between the closed conformity of the doctrine of progress founded in the scientific mastery of nature and the discontinuity or interruption of a failed or prophesied social insurrection. The text might productively be read in Benjaminian terms from the perspective of those destined to fail, left behind as anonymous witnesses in the narrative of an alternative history. The haunting opening of the poem, with its concatenation of 'dreary gleams', 'sandy tracts', and the shining of 'great Orion' and the Pleiades, gestures towards a kind of poetry which, in Hans-Georg Gadamer's terms, 'always transcends both poet and interpreter' in its pursuit of a meaning 'that points toward an open realm'. Such a realm comprises what Gadamer designates 'an effective whole in which everything described, the landscape and the dreaming I, is immersed and enveloped'.[21] Anne Janowitz interestingly suggests that the deployment of this type of 'landscape poetic' 'was helpful to Chartist poets insofar as it linked [...] the contemporary struggle to a communitarian past built in the countryside'.[22] In its inaugural moment Tennyson's poem is posited upon the self-reflexive trope of light and twilight, a figure which Gadamer has fruitfully elaborated:

> The light that causes everything to emerge in such a way that it is evident and comprehensible in itself is the light of the word. Thus the close relationship that exists between the shining forth of the beautiful and the evidentness of the understandable is based on the metaphysics of light.[23]

Such a philosophical interpretation, however, may be scientifically contextualized, since the Orion Nebula was, at the moment of the poem's genesis, the centre of an urgent scientific debate over what astronomers dubbed 'the dissolving view' – a debate keenly followed by the young Tennyson. John Herschel's series of sketches of the Orion Nebula, executed in the mid-1830s, made the phenomenon what Isobel Armstrong calls 'the obsessive test case for observation', the haze of the 'dissolving view' configuring 'a contradictory universe in which all elements were in a state of non-synchronic change'[24] – a state to which Tennyson's fevered narrator bears witness. Armstrong pertinently observes that, in an equally fevered mind, that of Thomas de Quincey reviewing an astronomical study in 1846, Orion served as 'a coded allegory of the results of inverting the order of things, and letting loose a primitive species – the working class – incapable of culture'.[25] The 'nebular hypothesis' – William Herschel's suggestion, following Kant and Laplace, that nebulae might be new sidereal systems or stars in the process of being formed – led to the phenomenon of evolving nebulae becoming the key emblem of astronomy as a progressively rational science.[26] Tennyson appears to refer to this as early as his Cambridge prize poem, 'Timbuctoo' (1829):

> [...] The clear Galaxy
> Shorn of its hoary lustre, wonderful,
> Distinct and vivid with sharp points of light,
> Blaze within blaze, an unimagin'd depth
> And harmony of planet-girded suns
> And moon-encircled planets, wheel in wheel,
> Arch'd the wan sapphire.[27]

As Anna Henchman observes, the idea 'that the universe was not inherently stable, but existed in a state of constant flux, was one of the most radical implications of stellar astronomy'. Such a theory, she adds, was instrumental in 'proposing that solar systems such as ours were derived from fluid bodies of gas and matter'.[28] According to William Herschel, in Pamela Gossin's account,

> the life cycle of nebulae begins when the largest star within nebulous clouds attracts others to it to form a cluster, or island universe, through the joint action of inward attraction and projective forces.[29]

Herschel's astronomical papers of the late eighteenth century suggested an evolutionary and expansionist model for the universe but also hinted at the possibility that the star system could one day wither away into a 'dark centre'. Henchman appositely notes that one of the putative results of the hypothesis was 'that the sun would eventually burn itself out'.[30] In his study of the scientific elements of Tennyson's poetry, M. Millhauser notes that 'The idea that the sun must eventually cool off was implicit in the nebular hypothesis, which held that the earth was originally part of its substance, but cooled more rapidly because of its smaller size.'[31] These prognostications led to increasingly atheistical interpretations, especially in France, which tended to discountenance the literal truth of Genesis. William Herschel's theory began with diffuse clouds of nebulosity which would eventually condense into star clusters, and late in his career he came to recognize that the Orion Nebula was situated within our own galaxy, whilst his son John provided, in his *Philosophical Transactions* (1833), an authoritative catalogue of over two thousand nebular clusters. One crucial issue, Henchman observes, 'centred on the nature of what appeared to be patches of gaseous matter in bodies like the nebula of Orion'.[32] The original 1832 version of 'The Palace of Art' had echoed these astronomical speculations: in the original, the female 'soul' scans the heavens with 'optic glasses' to observe

> Regions of lucid matter taking forms,
> Brushes of fire, hazy gleams,
> Clusters and beds of worlds, and bee-like swarms
> Of suns, and starry streams.[33]

And, like the 'Locksley Hall' protagonist, she is especially struck by 'the marvellous round of milky light / Below Orion', a theme strikingly elaborated by one of the female speakers in *The Princess*, who says,

> 'This world was once a fluid haze of light,
> Till toward the centre set the starry tides,
> And eddied into suns, that wheeling cast
> The planets:'[34]

Tennyson's verse bears witness to the symptomatic and evolutionary resonance of the discovery of star clusters and what, in his later astronomical novel, *Two on a Tower* (1882), Hardy would refer to as 'fine fogs, floating nuclei, globes that flew in groups'.[35]

To conclude: the 'shining forth' promised by science and evolutionary narratives disintegrates in Tennyson's poem, under the impress of the social crisis of early Victorian England, giving way to the 'dreary gleams' and 'vapour from the margin' which hint at auratic loss. The hero of *Maud*, significantly, will bear witness to the way in which he, listening to 'the tide in its broad-flung shipwrecking roar' (98), 'Walk'd in a wintry wind by a ghastly glimmer, and found / The shining daffodil dead, and Orion low in his grave' (100–101). If, in 'Locksley Hall' and subsequently in *Maud*, Tennyson may be defined as a poet 'in a destitute time' of scientific rationality, then as Heidegger postulates of Rilke,

> […] only his poetry answers the question to what end he is a poet, whither his song is bound, where the poet belongs in the destiny of the world's night. That destiny decides what remains fateful within this poetry. (*PLT*, 139)

Notes

1. J. F. C. Harrison, *The Early Victorians* (London: Weidenfeld and Nicolson, 1971), 12.
2. Alfred Tennyson, 'Locksley Hall', in *Tennyson: A Selected Edition*, ed. Christopher Ricks (Harlow: Pearson Longman, 2007). Subsequent references are to this edition.
3. Martin Heidegger, *Poetry, Language, Thought*, trans. A. Hofstadter (New York: Harper and Row, 1971), 90. Subsequently cited in the text as *PLT*.
4. Catherine Hall, *The Industrial Reformation of English Fiction, 1832–1867* (Chicago: University of Chicago Press, 1985), 114.
5. Kirstie Blair, 'Tennyson and the Victorian Working-Class Poets', in *Tennyson Among the Poets*, ed. R. Douglas-Fairhurst and S. Perry (Oxford: Oxford University Press, 2009), 294.
6. On the parallels between this passage and Saint-Simonian social theory, see John Killham, *Tennyson and 'The Princess'* (London: Athlone Press, 1958), 36–8.
7. Alan Swingewood, *Marx and Modern Social Theory* (London: Macmillan, 1975), 68.
8. Anne Janowitz, *Lyric and Labour in the Romantic Tradition* (Cambridge: Cambridge University Press, 1998), 144.
9. Patrick Joyce, *Visions of the People* (Cambridge: Cambridge University Press, 1991), 29, 336.
10. Homi Bhabha, *The Location of Culture* (London: Routledge, 1994), 130.
11. Patrick Brantlinger, *Rule of Darkness* (Ithaca: Cornell University Press, 1988), 9.
12. T. W. Adorno, *Minima Moralia*, trans. E. Jephcott (London: New Left Books, 1974), 156.
13. Ibid., 23.
14. Charles Kingsley, review in *Fraser's Magazine*, 1850, in *Tennyson: The Critical Heritage*, ed. J. D. Jump (London: Routledge and Kegan Paul, 1967), 179.
15. August Strindberg, preface to *Miss Julie*, in *The Father, Miss Julie, and the Ghost Sonata*, trans. M. Meyer (London: Eyre Methuen, 1976), 95.
16. Brantlinger, *Rule of Darkness*, 36.

17 Janowitz, *Lyric and Labour*, 143.
18 Walter Benjamin, *The Arcades Project*, trans. H. Eiland and K. McLoughlin (Cambridge, MA: Belknap Press, 2002), 473. Subsequently cited in the text as *AP*.
19 T. W. Adorno, *Aesthetic Theory*, trans. R. Hullot-Kentor (London: Continuum, 2004), 107. Subsequently cited in the text as *AT*.
20 Michael Sanders, 'Poetic Agency: Metonymy and Metaphor in Chartist Poetry', *Victorian Poetry* 39 (2007): 114.
21 Hans-Georg Gadamer, *The Relevance of the Beautiful*, trans. N. Walker (Cambridge: Cambridge University Press, 1986), 162.
22 Janowitz, *Lyric and Labour*, 157.
23 Hans-Georg Gadamer, *Truth and Method*, trans. J. Weinsheimer and D. G. Marshall (New York: Continuum, 1998), 483.
24 Isobel Armstrong, *Victorian Glassworlds: Glass Culture and the Imagination, 1830–80* (Oxford: Oxford University Press, 2008), 305, 309.
25 Ibid., 310. De Quincey was reviewing John Pringle Nichol's *Contemplations on the Solar System* (1844). It was Pringle's earlier *Views of the Architecture of the Heavens* (1837) which helped to establish the significance of the nebular hypothesis. Tennyson owned a copy of this volume, along with John Herschel's *Discourse on Natural Philosophy* (1830) and Mary Somerville's *On the Connexion of the Physical Sciences* (1835).
26 On this issue see James A. Secord, *Victorian Sensation* (Chicago: University of Chicago Press, 2000). Literary allusions to the nebular hypothesis were not uniformly portentous, as witness the process 'analogous to that of alleged formations of the universe' in *Far From the Madding Crowd*, when Bathsheba attempts to hive the bees and observes that the 'bustling swarm had swept the sky in a scattered and uniform haze, which now thickened to a nebulous centre'. Thomas Hardy, *Far from the Madding Crowd* (1874), ed. S. Falck-Yi (Oxford: Oxford University Press, 2002), 178.
27 'Timbuctoo,' in *Alfred Tennyson: The Major Works*, ed. A. Roberts (Oxford: Oxford University Press, 2009), 5.
28 Anna Henchman, '"The Globe We Groan In": Astronomical Distance and Stellar Decay in *In Memoriam*', *Victorian Poetry* 41 (2003): 33.
29 Pamela Gossin, *Thomas Hardy's Novel Universe* (Aldershot: Ashgate, 2007), 142.
30 Henchman, '"The Globe We Groan In"', 33.
31 M. Millhauser, *Fire and Ice* (Lincoln: Tennyson Research Centre, 1971), 19.
32 Ibid., 35.
33 'The Palace of Art' in *Tennyson: A Selected Edition*, 64.
34 *The Princess*, vol. 2, lines 101–104 in *Tennyson, A Selected Edition*, 243.
35 Thomas Hardy, *Two on a Tower*, ed. S. Ahmad (Oxford: Oxford University Press, 1993), 268.

Chapter 2

'TENNYSON'S DRIFT': EVOLUTION IN *THE PRINCESS*

Rebecca Stott

When the Darwinian naturalist T. H. Huxley described Tennyson as 'the first poet since Lucretius who has understood the drift of science',[1] he meant of course to compliment the poet on his ability to interpret science, to divine its general direction, but he might have got by without using the word 'drift'. As Huxley was a literary man, widely read and sensitive to language, we can assume the word was thoughtfully chosen. Drift means, at least as Huxley uses it here, 'the meaning, tenor, purport and scope' of science. Interestingly, it has an ambiguity of agency at its heart for it can be used to mean either *conscious* direction or action (as in 'What is your drift?' or 'Do you catch my drift?') or a movement driven randomly by natural or *unconscious* forces (as in a 'drift of leaves' or a 'drift of smoke'). Tennyson did indeed understand the drift of nineteenth-century science in ways that were broad, philosophical and insightful, and he played an important part in interpreting the new discoveries of science for a wide and trusting readership, but he also seems to have been curious about the ways in which the meanings of science were *made*. As an experiment in dialogic or conversational form, his long narrative poem of 1847, *The Princess: A Medley*, seeks, I will argue, to persuade us that educated *mixed-sex* conversation is the force that determines and shapes the drift of science.

***

The Princess was a long time brewing. Tennyson was 30 years old when he first conceived of the idea in 1839 and nearly 40 when he published the poem in 1847. Its gestation spans a decade, although the bulk of it was written between 1845 and 1847. It was, like several of the most important poems of the era, an experiment with conversational and narrative form, an attempt to use the poem not only to tell a story, but in this case to dramatize a series of contemporary debates about politics, gender and education. To the poem Tennyson brought his experience of education at Cambridge in the 1830s, his lifelong fascination with science and a still unresolved set of questions about women, education and science. The poem's almost 10-year gestation is, I believe, part of the reason it vacillates so strongly in the ideas and ideologies it expresses, part of the reason it is so difficult to read. Tennyson read widely during this time and roamed widely across many different ideas in conversation with others; his opinions on education, the role of women, politics and the origin of man were evolving during this time. The poem dramatizes, opens up and explores many of those ideas.

Tennyson was always fascinated by science. His son tells us in his memoir that the boy Tennyson, growing up in that crowded rectory full of simmering tensions and conflicts, spent long hours thinking about the stars, their origins and the beginnings of time; he scoured the newspapers for information about new stars or comets; he wrote poetry describing the surface of the moon in his notebooks, fractured lines interspersed with astronomical diagrams. Slowly he began to turn his scientific astonishments into poetry: 'The rays of many a rolling central star', he wrote in two of his very earliest lines of poetry, 'Aye flashing earthwards, have not reached us yet.'[2] He was also fascinated by the idea of deep time and curious about the long effects of natural processes on the landscape. 'From his childhood', Hallam Tennyson wrote, 'my father had a passion for the sea, and especially for the North Sea in wild weather – "the hollow ocean-ridges roaring into cataracts" [...] The cottage to which the family resorted [in Mablethorpe] was close under the sea bank, "the long low line of tussocked dunes". "I used to stand on this sand-built ridge", my father said, "and think that it was the spine-bone of the world." From the top of this, the immense sweep of marsh inland and the weird strangeness of the place greatly moved him.'[3]

When Tennyson took his place at Cambridge in 1827 he was no longer on his own with his pulse-racing speculations about the long history of the planets, the earth and species. All around him young men not only read widely across the sciences but also talked and argued freely about the implications of the startling new discoveries in physiognomy, geology and comparative anatomy, honing their rhetorical skills, testing the premises of their faith, challenging each other, searching for meanings and for narratives that tied all the new discoveries together into a coherent explanation of the earth's long history or that squared with the Biblical account of creation. These young men kept up with new European discoveries by reading the short reviews of new papers and books published in the *Quarterly Review* and the *Westminster Review*.[4]

In 1828 Tennyson read a review in the *Quarterly* that shocked and excited him: it described new advancements in the understanding of the nervous system and included an account of Friedrich Tiedemann's remarkable discovery that in its stages of development the brain of the human foetus closely resembled the brains of other vertebrates like fishes, reptiles, birds and the lower Mammalia.[5] For most Cambridge undergraduates who had more than a passing interest in the new sciences, like Tennyson, Tiedemann's discovery would have been taken as further proof of the controversial theory proposed by several French comparative anatomists and zoologists – either that all animals, large and small, shared a common archetype or blueprint or that they had evolved from simple single-celled aquatic organisms in a primal sea millions of years earlier, passing infinitely slowly through stages of increasing complexity and diversification. Jean-Baptiste Lamarck, Professor of Invertebrates at the Museum of Natural History in Paris, had been arguing since 1802 that the first simple aquatic organisms had become fish, then lizards, then birds, then eventually humans. It seemed, from Tiedemann's work, that the proof and marks of that proposed – and highly controversial – history lay in the foetal human brain.

Over the following weeks Tennyson exchanged letters with his friend Arthur Hallam about the theological implications of Tiedemann's discovery, asking questions about the nature of the soul.[6] Soon afterwards he summarized the ideas at the heart of the review at a meeting of the Cambridge Apostles, suggesting that Tiedemann's work proved that man had evolved from simpler organisms: 'My father', wrote Hallam Tennyson, 'seems to have propounded in some college discussion the theory that the development of the human body might possibly be traced from the radiated, vermicular, molluscous and vertebrate organisms.'[7] Soon after, Tennyson began to explore the ontological and social ramifications of the idea, excitedly, in a poem still in formation, 'The Palace of Art', first published a few years later in 1832. Here a personified Soul reveals her secrets:

> 'From change to change four times within the womb
> The brain is moulded,' she began,
> 'So through all phases of all thought I come
> Into the perfect man.
>
> 'All nature widens upward. Evermore
> The simpler essence lower lies,
> More complex is more perfect, owing more
> Discourse, more widely wise.'
> (141–8)

The lines, written probably between 1828 and 1830, mark the beginning of Tennyson's engagement with what we might call the metaphysics of evolutionary speculation; he attempts to synthesize the fragments of new discoveries in diverse fields into a large-scale narrative about the history of the earth that will also serve to explain its present, its future and the effects of time on future races, and in particular help define the place of progress and reform in society. Tennyson saw a role for himself in striving to interpret and explain the implications of these new sciences at a time when British science was, in response to a much more speculative French science, defining itself as rigidly empiricist, concerned only with the slow and deliberate accumulation of facts.

Conversations about the supposed evolution of species, in Cambridge in particular but also elsewhere in the country, were regarded in the 1830s and '40s not only as dangerously atheist – or to use a nineteenth-century term, dangerously 'infidel' – in that they challenged Biblical accounts of creation, but also as largely 'French'. (To many this meant unpoliced, infectious, subversive, anticlerical and prone to provoke revolution and convulsive disorder.) In the '30s and '40s in particular, transmutation (or the development theory, as such ideas were called in Britain in this period) was also often bound up with radical or reformist discussions about the progress and future of humankind, and occasionally specifically about womankind.[8] Jean-Baptiste Lamarck's concept of evolution – that animals can improve themselves by their own efforts and pass on their traits to their offspring – seemed to prove that all organisms were equally capable of advancement and that a system that kept them in a closely policed, divinely-ordained social hierarchy worked against nature.

This was a period of political speculation in Britain and in this climate, between the first and second reform bills, and in the shadow of the French Revolution, evolutionary ideas, and particularly those of Lamarck, were widely used by radicals such as Paineites, Saint-Simonians and Owenites to underpin a reformist agenda and to undermine the power and authority of the church.[9] Thus, evolutionary ideas in these decades before the publication of Darwin's *Origin* were not just controversial because they contravened Biblical accounts; they were controversial because they had the potential to be politically and socially subversive.

***

If we glance sideways at Darwin in these same years, his first encounters with evolutionary ideas show a striking similarity to those of Tennyson. When he was growing up in Shrewsbury, Darwin's scientific interests were more concrete than Tennyson's and on a smaller scale – he was a collector of small things: shells, birds' nests, stamps and minerals. When he and his older brother Erasmus were home from school they conducted chemical experiments in the garden shed, which they equipped with test tubes, stopcocks, crucibles, retorts, evaporating dishes and burners. But for Darwin, like Tennyson, science only came fully alive in speculative conversation, only became fully astonishing in the debates held by all-male student natural history societies loosely connected to Edinburgh Medical School, where he enrolled as a student in 1825 at the age of 16. Here he met like-minded young men who were engaged in scientific pursuits and hobbies and followed the latest developments in physiology, geology and comparative anatomy in France and Germany, men who wanted to debate the philosophical, political and theological implications of these developments.

In Edinburgh, either in one of the student societies or on the beach at Leith where he collected sea creatures for his experiments, Darwin met an older man called Robert Grant, a local doctor and one of the most remarkable and original men of science in Scotland; Grant's ideas, he remembered later, truly *astonished* him. Robert Grant, an expert on sea sponges as well as doctor, had determined to recruit the boy stranger, seen so often on the beach in the past year, as an additional assistant in his marine experiments. To his amazement he discovered that the boy was the grandson of the great Erasmus Darwin. He clearly did not understand how important his grandfather's great book, *Zoonomia*, had been to the advance of science or how bold it had been in advancing evolutionary ideas.

Conversations with Grant about the evolutionary ideas of the radical French professor Jean-Baptiste Lamarck, who had taught Grant in Paris, and those of Darwin's grandfather Erasmus Darwin, unfolded for the boy Darwin on the beach in Leith, and they continued for eighteen months as Grant, Darwin and Grant's assistant John Coldstream walked and worked the shoreline of the Firth of Forth, or talked excitedly back in Grant's house in the seaside village of Prestonpans, or gave papers, or compared notes, or opened up zoophytes and watched zoophyte eggs swim, or joined in the heated debates at the student societies. These conversations with Grant and with fellow students – which ranged from facts to their interpretation and particularly concerned the political, social and theological meanings of the new discoveries – resonated in Darwin's head for the rest of his life. Grant and Darwin eventually quarrelled, and then their lives took them in different directions. Grant moved to London to take up a prestigious post as Professor of Zoology at the newly instituted University of London. Darwin left for Cambridge and eventually sailed on the *Beagle* in 1831 as ship's naturalist, where for four years he continued not only to put together an extraordinary collection of natural history specimens but also to test out Grant's ideas about the origins of species.[10]

***

Darwin and Tennyson entered these scientific debates when they were still in formation and at a time when no one had yet attempted to synthesize all the disparate discoveries – from zoology, geology, palaeontology – into a single narrative; to do so in the established atmosphere of rigid British empiricism would have been risky. Then, first in 1830–32, and secondly in 1844, two groundbreaking but very different British books appeared that attempted to forge a grand narrative from the new and still emerging discoveries. The second of the two brought scientific speculation about origins to a much larger and increasingly mixed-*sex* audience. Both profoundly shaped the future direction of Darwin and Tennyson's writing.

The first, a book by Charles Lyell called *Principles of Geology* (in three volumes, published in 1830, 1832 and 1833), brilliantly synthesized a series of new geological discoveries into a narrative about the formation of the earth since its beginnings. Lyell explained how landscapes had formed infinitely slowly and over eons of time through natural forces still in operation (tides, seas, wind, rain, river and ice), and he painted a picture of an earth that had been, since its beginnings in deep time, in a state of continual movement and metamorphosis and immeasurably older than had yet been admitted. However, whilst Lyell believed that the earth's crust had shifted continuously and continued to do so, he passionately refuted the idea of species change. Over the course of 40 pages in the second volume, he presented and refuted Jean-Baptiste Lamarck's evolutionary claims in minute detail. In refuting Lamarck's progressive natural processes, Lyell instead stressed nature's destructions and derelictions, portraying organisms pitted against each other for survival: 'Every species', he wrote, 'which has spread itself from a small point over a wide area, must, in like manner, have marked its progress by the diminution, or entire extirpation, of some other, and must maintain its ground by a successful struggle against the encroachments of other plants and animals.'[11] Lyell's was a dark and pitiless Nature.

Darwin read all three volumes of *Principles* on the *Beagle* in 1832–34, taking the first volume with him, collecting the second in Montevideo in 1832 and the third in Valparaiso in 1834, and he remembered later that much of what he witnessed on his travels he saw excitedly through Lyell's eyes. Tennyson, his son tells us, was 'deeply immersed' in *Principles* through much of 1837.[12] For Tennyson, until now excited and encouraged by Lamarck's thesis that all species were moving forward progressively, Lyell had thus 'destroyed a dream only to substitute a nightmare'[13] by presenting a vision of a chaotic and uninterested Nature, moving not in straight lines towards an ever-improving future but working instead through cycling extinctions and conflicts.[14]

The second of these two groundbreaking books was Robert Chambers's sensational and bestselling *Vestiges of the Natural History of Creation* published anonymously in November 1844. Beautifully written and grippingly told, the book brought together new discoveries in physiognomy, geology, embryology and comparative anatomy to create another kind of biography of the earth that included mutating species as well as landscapes and that was resolutely optimistic in contrast to Lyell's *Principles*. The stars, planets and moons had evolved from a gaseous 'Fire-mist', the author claimed confidently; minute invertebrate animals had formed in the water that covered the cooling earth; these primitive forms had evolved over millions of years into fishes, amphibians, reptiles, mammals and finally man. The author described this process as the 'universal gestation of nature' – everything, rocks and bodies, were evolving and continued to do so. The narrative voice was spritely, confident and – most importantly – reassuring. Readers were reassured that they had no reason to doubt that there was a guiding hand behind what seemed like divine indifference: 'There is a system of Mercy and Grace behind the screen of nature, which is to make up for all casualties endured here, and the very largeness of which is what makes these casualties a matter of indifference to God. [...] it is necessary to suppose that the present system is but a part of a whole, a state in a Great Progress'.[15]

Vestiges gave a sensational narrative – albeit deeply controversial and in some ways factually flawed – to the earth and to man's place on it; it also importantly shifted the timbre and tone of evolutionary speculation into a more optimistic phase which insisted on progress and improvement as Nature's way: God, the author proposed, had made the earth and had given it natural laws which had worked their way through for millions of years. Progress was evident everywhere, the author claimed, but although God did not intervene in those laws, he was still present at a distance.

In mid-November 1844, whilst he was writing both *In Memoriam* and *The Princess*, Tennyson read a lengthy lead review of *Vestiges* in the *Examiner*, a weekly radical reform paper, and immediately wrote to his bookseller to ask him to send him a copy. 'It seems', he wrote excitedly, 'to contain many speculations with which I have been familiar for years, and on which I have written more than one poem.'[16] He was right to act quickly because the first edition of the book sold out in a few days.

As Tennyson turned the pages of *Vestiges* and contemplated its extraordinary claims, the book became the talk of the season, and controversial science came into mixed-sex conversation across Britain.[17] Whilst Lyell's *Principles* had been written for a specialized scientific audience, *Vestiges* was written for the educated general reader and published at a price that made it affordable to a large sector of the literate population. Benjamin Disraeli wrote to his sister, '*Vestiges* is convulsing the world, anonymous,' and Mary Disraeli wrote, 'Dizzy says it does & will cause the greatest sensation & confusion.'[18]

Vestiges became the talk of dinner tables, but, written for a large audience in accessible and elegant prose, it also became the subject of debate and correspondence between men and women. The small red-cloth volume galvanized thousands of male and female readers into taking positions on a range of speculative – and potentially heretical – ideas about the origins of life, the role of women in society, the future of the race, human kinship with animals, the nature of the soul, the age of the earth and the nature and existence of God. It provoked readers into forming opinions; it compelled them to assemble evidence from the pages of the book and to form arguments and counterarguments. The 24-year-old Florence Nightingale, for instance, attended a dinner in February 1845 at Sir William Heathcote's country residence. The guests, she remembered, discussed mesmerism, Boston, and then *Vestiges*. She recorded in her diary that 'when we parted, we had got up so high into *Vestiges* that I could not get down again, and was obliged to go off as an angel.'[19] Later that year, once she had assimilated the full philosophical and political import of the book, she wrote after attending another dinner party: 'Is all that china, linen, glass necessary to make man a Progressive animal?'[20] Nightingale's reaction to the book is typical of many of the time: initially she could only see a kind of grotesque comedy in the idea of mutating and transforming body parts and animal–human kinship, but later she came to see how evolutionary ideas might profoundly challenge fixed social hierarchies.[21]

What do we know about Tennyson's experience of mixed-sex conversation during the mid 1840s when he was writing *The Princess*? Only glimpses. He was living in London for much of the time and, when not taking the water cure for a range of anxieties and health problems, he was nursing his long grief at the loss of Hallam and new grief at the suspension of his engagement to Emily Sellwood. He was also gradually shedding his natural shyness by attending small informal missed-sex social gatherings at the homes of close married friends such as the Howitts and the Carlyles, often presided over or attended by strong-minded, clever and widely-read women such as Jane Welsh Carlyle. For Tennyson, the days of all-male student society conversation were giving way to the mixed-sex dinner party conversation of young married London households. During April and May 1845, according to his biographer Robert Bernard Martin, 'the journal of Aubrey de Vere and the letters of Brookfield, to name only two sources, list evening after evening when Tennyson was in company. [...] In such small groups Tennyson's conversational brilliance and charm came out'.[22]

It seems from first-hand accounts of these events that in the mid-1840s, the protocols of dinner party behaviour and mixed-sex social conversation were becoming more elastic in these London circles, to the degree that Tennyson was not always sure how to behave. Protocols were sufficiently relaxed at the Carlyles' house in Chelsea, for instance, for Tennyson to feel he could turn up unannounced. Jane Carlyle describes one such occasion in January 1845:

> [I] had made up my mind for a nice long quiet evening of looking into the fire, when I heard a carriage drive up, and men's voices asking questions, and then the carriage was sent away! And the men proved to be Alfred Tennyson of all people and his friend Mr Moxon – Alfred lives in the country and only comes to London rarely and for a few days so that I was overwhelmed with the sense of Carlyle's misfortune in having missed the man he likes best. [...] Alfred is dreadfully embarrassed with women alone – for he entertains at one and the same moment a feeling of almost adoration for them and an ineffable contempt! Adoration I suppose for what they might be – contempt for what they *are*! The only chance of my getting any right good of him was to make him forget my womanness – so I did just as Carlyle would have done had he been there; got out pipes and TOBACCO – and brandy and water – with a deluge of tea over and above, – The effect of these accessories was miraculous – he professed to be ashamed of polluting my room 'felt' he said 'as if he were stealing cups and sacred vessels in the Temple' but he smoked on all the same – for three mortal hours! – talking like an angel – only exactly as if he were talking with a clever man – which –being a thing I am not used to – men always adapting their conversation to what they take to be a woman's taste – strained me to a terrible pitch of intellectuality – When Carlyle came home at Twelve and found me all alone in an atmosphere of tobacco so thick that you might have cut it with a knife his astonishment was considerable![23]

Jane Brookfield, another acquaintance of Tennyson's during these years, newly married in the 1840s to William Henry Brookfield, Tennyson's friend from Trinity, and who ran a successful London salon, describes an occasion in which Tennyson expressed his frustration at social protocols:

> At one time, he told me he very much wished to find out whether ladies liked their male acquaintances to assume a gentler tone of voice, when speaking to them, from that in which they talked to each other. Alfred said he disliked this affectation of consideration towards what is called 'the weaker sex', and that he preferred to think that the tone of voice, as well as the subject of conversation, should need no remodeling to make it fit for ladies to hear.[24]

This was the season in which *Vestiges* dominated dinner party conversations across the country, so – whilst we do not *know* that Jane Carlyle, Moxon and

Tennyson discussed the controversial book in that haze of pipe tobacco in January 1845 – given that the book had only been published in October of the previous year and that Tennyson had just read it, and that just about everybody in London intellectual circles was reading it – including Prince Albert who was reading it aloud to Queen Victoria in the afternoons – it is probable that they did do so. (By August 1845 Jane was complaining in another letter of the 'eternal Vestiges' and of a male bore holding forth at a dinner party about its significance.[25]) A large number of the descriptions of discussions of *Vestiges* that took place in the 1840s, carefully gathered by James A. Secord in *Victorian Sensation*, indicate that these discussions were often of a mixed-sex nature.

Vestiges was the first science book of the decade, almost certainly of the year, to be discussed widely by both men and women, and upon which women formed and voiced sophisticated opinions. Benjamin Disraeli must have heard a great number of women express their opinions on *Vestiges* before he decided to use his novel *Tancred* to parody them. In the novel Lady Constance tries to persuade the hero to read *Vestiges* (here called *The Revelations of Chaos*); Disraeli mocks Lady Constance's breathless evangelical excitement:

> After making herself very agreeable, Lady Constance took up a book which was at hand, and said, 'Do you know this?' And Tancred, opening a volume which he had never seen, and then turning to its title-page, found it was 'The Revelations of Chaos,' a startling work, just published, and of which a rumour had reached him. 'No,' he replied; 'I have not seen it.'
>
> 'I will lend it you, if you like. It is one of those books one must read. It explains everything, and is written in a very agreeable style.'
>
> 'It explains everything?' said Tancred. 'It must, indeed, be a very remarkable book!'
>
> 'I think it will just suit you,' said Lady Constance. 'Do you know, I thought so several times while I was reading it?'
>
> 'To judge from the title, the subject is rather obscure,' said Tancred.
>
> 'No longer so,' said Lady Constance. 'It is treated scientifically; everything is explained by geology and astronomy, and in that way. It shows you exactly how a star is formed. Nothing can be so pretty! A cluster of vapour – the cream of the milky way – a sort of celestial cheese-churned into light. You must read it – 'tis charming.'
>
> 'Nobody ever saw a star formed,' said Tancred.
>
> 'Perhaps not. You must read the "Revelations." It is all explained. But what is most interesting is the way in which man has been developed. You know, all is development. The principle is perpetually going on. First, there was nothing; then there was something; then – I forget the next – I think there were shells, then fishes; then we came – let me see – did we come next? Never mind that – we came at last. And the next change there will be something very superior to

us – something with wings. Ah! that's it. We were fishes, and I believe we shall be crows. But you must read it.'

'I do not believe I ever was a fish,' said Tancred.[26]

In the wake of the sensational talk that *Vestiges* made, Tennyson and Darwin, poet and naturalist, now turned in quite different directions. Darwin, horrified at the opprobrium being meted out to the book's still-anonymous author, turned to the small scale and the lowborn, determining to answer the riddle of a single aberrant barnacle he had collected on the *Beagle*. The publication of intricate and accurate taxonomic work, he decided, was the only way to earn his spurs before he dared to publish his species work.[27] The result was a myopic, fact-freighted, descriptive and resolutely unspeculative work: a series of four volumes that chronicled the minute differences between barnacle species.[28] During the eight-year empirical research, he frequently struggled to repress what he called the 'speculatist' in himself, the man who was frustrated by the injunction to be a systematist above all else. Speculation simply had no place in the work he had undertaken.[29]

Tennyson turned instead to the large scale, the speculative, the highborn and the fantastic, to an unfinished poem about a princess, at that point tentatively called 'The New University'. He had glimpsed in *Vestiges* a potential resolution to the large-scale metaphysical, political and philosophical questions that had pressed upon him for more than a decade. The poem gave him an opportunity to synthesize his own ideas and reactions to the new scientific accounts of time and origins and speculate on what they might mean for the future of humankind. 'The New University' became, in the wake of Tennyson's reading of *Vestiges*, a formally experimental and risk-taking narrative poem called *The Princess: A Medley*, a poem made up of a number of voices in dissent and in debate and which was, in the words of critic Elaine Jordan, a poem 'in more than one mind.'[30]

The Princess, though conceived around 1839, was largely written whilst the controversy over *Vestiges* was at its peak, forged out of the mixed-sex conversations that dominated the talk of dinner tables in 1845 and 1846; Tennyson put the last touches to the poem in November 1847. The narrative tells the story of group of seven male undergraduates of Tennyson's time who attend a summer fete at a country house. They join a small group of women in the Gothic ruins in the grounds of the house and agree to tell a story by turns, mimicking a game that Tennyson had played with his friends at Cambridge. Together the seven male narrators compose a story about a Prince and his companions who have dressed as women in order to enter a women-only university run by a Princess to whom the Prince is betrothed by birth and to whom he wishes to press his suit. Over several days the Prince and Princess have several conversations about science, politics and the education of women whilst the Prince's disguise holds. When the Princess discovers his identity she banishes him. When the Prince's father and the Princess's brother turn up inclined to make a fight, a violent joust begins in which the Prince is wounded and falls into a coma. The Princess nurses the Prince back to life and falls slowly in love with him. He proposes again. The poem ends before she has given her answer.

Critics have discussed the ambiguity-riven sexual politics of *The Princess* in detail but have, until recently, rarely considered the medley of contending ideas drawn from geology, physiology and comparative anatomy at its heart, except to search out the exact source passages for the scientific ideas.[31] Two recent exceptions are fine essays by Michael Tomko and Virginia Zimmerman.[32] *The Princess* is, unsurprisingly for Tennyson, a poem that seems divided against itself, in 'more than one mind'. Eve Kosofsky Sedgwick declares that the poem wants to pass itself off as enlightened in its sexual politics but is actually deeply concerned with the eroticized homo-social bonding of men.[33] Alan Sinfield argues that whilst *The Princess* dabbles with enlightened feminist ideas and opens up a debate that expresses radical ideas in the heart of the poem, it nonetheless closes down its own potentially radical multi-voicedness, subsuming several voices into a single sentimental conservative voice at its end.[34]

In *The Princess* Tennyson attempts to engage with the metaphysics and politics of evolution at a very early stage in the history of its assimilation and interpretation. That experiment compelled him to reach for a form that incorporated conversation and appositional techniques in order to show how the 'drift of science' (its meanings and its imports) are negotiated and forged by young men and women in conversation. Given that Tennyson had witnessed the interpretation, meaning and synthesizing of new, fragmented and controversial scientific discoveries being made all around him, his attempt to render that making into poetry could only authentically have been a 'medley'; Tennyson gave this term to the poem as a subtitle, thus drawing attention to his resolution to make one narrative from a multi-voiced heterogeneity.[35]

It is possible that Tennyson explained the nature of the poem's 'medley' structure and its relation to the age to friends who later reviewed the poem. In a review in the *Edinburgh Review* in 1849 Aubrey de Vere praised the poem's structure as suiting the heterogeneous nature of the age and time, and he used terms that appear interestingly close to those Tennyson used himself both within the poem and outside it: 'If a man were to scrutinise the external features of our time', de Vere wrote, ' […] he would be tempted, we suspect, to give up the task before long, and to pronounce the age a Medley.' The heterogeneity of the age, de Vere argued, extended to philosophy, art, society, politics and architecture. 'In this respect, Mr Tennyson's poem

"*The Princess*", not without design if we may judge by the title, resembles the age.'[36] The poem offers, Charles Kingsley claimed in another review of 1850, 'a mirror of the nineteenth century, possessed of its own new art and science, its own new temptations and aspirations, and yet grounded on, and continually striving to reproduce, the forms and experiences of all past time'.[37] It was a unification project which was also shared by Robert Chambers's *Vestiges*; Chambers too gave himself the task of revealing the unity at the heart of apparent disunity. He wrote, for instance, in *Vestiges*:

> These facts clearly show how all the various organic forms of our world are bound up in one – how a fundamental unity pervades and embraces them all, collecting them, from the humblest lichen up to the highest mammifer [mammal], in one system, the whole creation of which must have depended upon one law or decree of the Almighty, though it did not all come forth at one time.[38]

The Princess seeks to render fragments into a whole and thus to find synthesis, but at the same time it appears to resist subsuming difference into unity. Its representation of evolutionary ideas is similarly multi-voiced and open-ended. If Tennyson concludes *In Memoriam* with the acceptance that nature is 'as an open book' (2884), *The Princess* dramatizes the range of nature's readings and the various means by which it can be read. In the famous evolutionary dialogue between the Prince and the Princess which takes place at a dramatic cataract where rushing water has exposed the bones of a dinosaur,[39] the Princess reads the open book of nature in order to challenge the Prince's outmoded and patriarchal ways of thinking about women. She uses the discoveries of the new sciences to underpin her educational social engineering scheme, her vision for the future of womankind and her passionate belief in progress. If the mighty fossilized bones of the cataract speak of the difference between barbarism and present kind, she tells the Prince, then what might a future woman be like – she *that will be*?

Evolution is politically utopian for the Princess, just as it was for Florence Nightingale, the Saint-Simonians and the young dissenting medical men of St Barts who entertained Lamarckian ideas, because it meant that nothing was fixed. The Princess is convinced that natural laws enhanced by social regulations and human laws will ensure the progressive improvement of the race. The Prince is sceptical. The two argue about the extent to which men and women can shape the future direction of the race, as they stand in the shadow of the dinosaur bones:

> She gazed awhile and said,
> 'As these rude bones to us, are we to her

That will be.'
'Dare we dream of that', I asked,
'Which wrought us, as the workman and his work,
That practice betters?'
(278–82)

[Earlier the Prince had asked what she will feel if her grand scheme were to fail:]

'Think;
Then comes the feebler heiress of your plan,
And takes and ruins all; and thus your pains
May only make that footprint on the sand
Which old-recurring waves of prejudice
Resmooth to nothing [...] '
(221–5)

[To which she replies:]

'Would, indeed, we had been,
In lieu of many mortal flies, a race
Of giants living, each, a thousand years,
That we might see our own work out, and watch
The sandy footprint harden into stone.'
(250–55)

The timbre of the rhetoric in these passages is impressive. It moves fast; both participants are philosophically adept; both seek truth; both know the philosophical terrain well. Every shift in the sequence of the poem, William David Shaw has observed, 'opens another window in the reader's mind. As in Symbolist poetry, meanings that defy translation can be left for the reader to supply for himself.'[40] The same might be said about the evolving conversation between the Prince and Princess. Readers are not just silent witnesses in these exchanges, but are philosophically engaged, provoked into taking positions in the range of shifting and negotiated interpretations of nature dramatized for them.

These evolutionary speculations are all, one feels, metaphysical, political and theological interpretations of geological and fossil evidence considered, shaped and developed by Tennyson himself in conversation with others. The poem returns again and again to Tennyson's preoccupations with the nature and power of time. When, in the Prologue, one of the narrators proposes that storytelling, particularly collective storytelling, developed as a means to survive the long dark nights of winter, Lilia immediately refers to Time as a tyrant that must be killed.

> 'Kill him now,
> The tyrant! kill him in the summer too,'
> Said Lilia; 'Why not now?' the maiden Aunt.
> 'Why not a summer's as a winter's tale?
> A tale for summer as befits the time,
> And something it should be to suit the place,'
> (197–206)

The ancient image of a benign 'Mother Nature' had been profoundly challenged by Lyell in his account of the derelictions, indifference and destructions of Nature in *Principles*.[41] Tennyson had expressed the profound darkness of Lyell's vision in his chilling portrait of a monstrous, cannibalistic Mother Nature in *In Memoriam* as 'Nature, red in tooth and claw',[42] but he had other contending images to hand too: here he brought on Time the male tyrant. Robert Chambers had sought to restore some of the supposedly benign maternal qualities of Nature in *Vestiges* by using a repeated image of the mutually dependent family with the mother at its centre as one of his central metaphors. The speakers in *The Princess* significantly move through a sequence of shifting personifications of time and relationships with it – from Lilia's expressed intention to 'kill' the tyrant Time at the beginning of the poem to the Princess's idea of the importance of 'shaping' and 'serving' Time, to the final image, expressed by the Prince, of Time as a benign, protected, skirted woman:

> And so these twain, upon the skirts of Time,
> Sit side by side, full-summ'd in all their powers,
> Dispensing harvest, sowing the To-be,
> Self-reverent each and reverencing each,
> Distinct in individualities,
> But like each other ev'n as those who love.
> (3110–14)

Time shifts gender in the poem. The transition from a tyrant who must be killed to a care-giving mother mirrors the movement of the Prince and the Princess as together they negotiate an optimistic, progressive and mutually enhancing theory of the workings of time. The transition marks a movement away from the Lyellian version of time, which dominates *In Memoriam*, towards the more benign version of time drawn from *Vestiges*, which dominates the final passages of *The Princess*.

The evolutionary poetics of *The Princess* is significantly different from that of *In Memoriam*. It is almost as if putting the two into dialogue creates something like the two halves of Shakespeare's *The Winter's Tale* – the first half of Shakespeare's play revolves around a tragic malaise and doubt, a gaze that is inward-looking and profoundly destructive, both individually and collectively; the second half counters this malaise with play, jouissance and metamorphic redemption. Tennyson's poem deliberately evokes Shakespeare's play both in its language and in its plot. Just as the Prince is nursed back into life after his supposed death and then coma, Hermione, shocked into a symbolic paralysis, is warmed back into life by the return of her daughter and the remorse of her husband.

But the poem is not just a sequence of ideas. It enacts its ideas as a lyrical, linguistic and aesthetic performance. The movement from heterogeneity to homogeneity, the Prince and Princess's negotiations and renegotiations of their relationship to each other, their interpretations of the book of nature and their evolving views about man's role in shaping time in educating the future race, are mirrored in the poem in a sequence of lyrical descriptions of metamorphoses and immersions, often eroticized. The Prince turns from a man to a woman to a man again. The Princess falls into the water and is rescued by the Prince. Both fall and fall again. In the lyrics of *The Princess* Tennyson celebrates physical merging as immersion or falling or furrowing or enfolding, one element penetrating another or being absorbed into another, mimicking the slow movement of the Prince and Princess towards union. The crimson petal lyric is the most famous and most erotic of these lyrics:

> Now sleeps the crimson petal, now the white;
> Nor waves the cypress in the palace walk;
> Nor winks the gold fin in the porphyry font:
> The fire-fly wakens: waken thou with me.
>
> Now droops the milkwhite peacock like a ghost,
> And like a ghost she glimmers on to me.
>
> Now lies the Earth all Danaë to the stars,
> And all thy heart lies open unto me.
>
> Now slides the silent meteor on, and leaves
> A shining furrow, as thy thoughts in me.
>
> Now folds the lily all her sweetness up,
> And slips into the bosom of the lake:

So fold thyself, my dearest, thou, and slip
Into my bosom and be lost in me.
(2999–3012)

As critics have pointed out, the poem describes a series of enfoldings, furrowings, slippages, openings and slidings ('with me [...] on to me [...] unto me [...] in me [...] lost in me'), but because the lyric is spoken by a woman it is never clear whether what is enfolding or being enfolded is male or female. The merging seems to work both ways. And that is of course the point, as is made clear in the Prince's final speech, which though it has been much derided by critics as false-sublime or as the summation of the Prince's ideological manipulation of the Princess, nonetheless embodies the eroticized male–female merging that is at the heart of the poem's celebrations of evolutionary processes. 'True marriage', the Prince proposes, makes full ('fulfils') halves that are incomplete in isolation. The merging does not result in homogeneity but in a strange plural-singular entity. The merging or fitting of 'thought in thought', 'purpose in purpose', 'will in will' in this speech results in neither homogeneity nor heterogeneity but instead creates a strange oscillation between the singular and the plural for: '*they* grow / The *single* pure and perfect animal' with the '*two*-celled heart' beating, / with *one* full stroke, / Life' (my italics).

'Dear, but let us type them now
In our own lives, and this proud watchword rest
Of equal; seeing either sex alone
Is half itself, and in true marriage lies
Nor equal, nor unequal: each fulfils
Defect in each, and always thought in thought,
Purpose in purpose, will in will, they grow,
The single pure and perfect animal,
The two-cell'd heart beating, with one full stroke,
Life.'
(3119–28)

Elaine Jordan has described *The Princess* as erotically charged; indeed she has described Tennyson as writing some of the most erotic poetry since the seventeenth century.[43] But the undeniable eroticism of *The Princess* serves an ideological purpose as well as an aesthetic one; not only does it celebrate union and romanticize the Prince's quest, it also eroticizes change. If, as the Princess seeks to persuade the Prince, everything is in a state of perpetual, infinitely slow transformation and nothing is static or fixed, if slippage is the state of nature, then sex is the means by which new future comes into being.

Finally, in returning his readers to the outer frame of the story, to the undergraduates and their female friends sitting in the Gothic ruins (Lilia still angry at the position of women in the world, still unappeased), Tennyson refuses to allow his readers to remain in the fantasized past he has created; he insists that they now consider what they have read in relation to the present day. Ida's final speech, in which she accepts change, rings through the final lines of the poem, and in returning to the present day Tennyson asks his reader to see the change that the poem advocates, eroticizes and naturalizes, as clearly political. Ida declares to the Prince it might be a kind of manifesto:

> 'But trim our sails, and let old bygones be,
> While down the streams that float us each and all
> To the issue, goes, like glittering bergs of ice,
> Throne after throne, and molten on the waste
> Becomes a cloud: for all things serve their time
> Toward that great year of equal mights and rights,
> Nor would I fight with iron laws, in the end
> Found golden: let the past be past; let be
> Their cancelled Babels: though the rough kex break
> The starred mosaic, and the beard-blown goat
> Hang on the shaft, and the wild figtree split
> Their monstrous idols, care not while we hear
> A trumpet in the distance pealing news
> Of better, and Hope, a poising eagle, burns
> Above the unrisen morrow:'
> (1385–99)

And here in the final outer frame, Tennyson reminds us that the moment of the poem is one of 'revolts, republics and revolutions'. In discussing the threat of revolution that many feared would spread from Europe to Britain, the narrator implies that Britain has in such a moment to make choices: as an island she can either keep herself aloof, like the Princess, remain outside the movements of time or political processes, or 'serve' Time, 'shape' Time, and thus commit to reform and commit especially to the education of the future race, an education that must include working men and women. This is the only way to ensure social progress:

> 'Have patience,' I replied, 'ourselves are full
> Of social wrong; and maybe wildest dreams
> Are but the needful preludes of the truth:

For me, the genial day, the happy crowd,
The sport half-science, fill me with a faith.
This fine old world of ours is but a child
Yet in the go-cart. Patience! Give it time
To learn its limbs: there is a hand that guides.'
(3255–62)

In this same decade, whilst Tennyson strove to work out the full political and social implications of a full acceptance of evolutionary ideas, Darwin tried to erase the French – and thus political – associations from his species theory by grounding his work in the ordinary, the ubiquitous, the British and the commonplace: earthworms, bees, pigeons and barnacles. Robert Grant and Grant's unapologetically political ideas about evolution continued to act as a thorn in the side of the Anglican establishment in London and as a thorn in Darwin's side. When he later wrote the historical sketch to the *Origin*, Darwin worked hard to make sure there were as many ordinary British vicars and naturalists on his list of evolutionary predecessors as there were Frenchmen. Evolution, he knew, must be made safe and respectable and dissociated from French philosophy before anyone would be prepared to entertain it.

Tennyson took a different path. He did not seek to depoliticize evolutionary speculation or to separate out the new discoveries from their potential social or political meanings. Transmutationist ideas fascinated him; they opened up new justifications for reform, for the education of working men and of women and, once he had read *Vestiges*, seemed also to offer a hopeful future for mankind. Tennyson sought to do what *Vestiges* had done for its readers, to reassure them (and himself) that it was possible to embrace these new ideas, to interpret them, and to see hope in them, and for the sky not to fall nor for men with pitchforks to rise on the streets of London. It was a project of reassurance that had profound similarities with Chambers's *Vestiges*:

> And in this faith we may well rest at ease, even though life should have been to us but a protracted disease, or though every hope we had built on the secular materials with our reach were felt to be melting from our grasp. Thinking of all the contingencies of this world as to be in time melted into or lost into the greater system, to which the present is only subsidiary, let us wait the end with patience, and be of good cheer.[44]

Like Shakespeare's last plays, on which it is in part modelled, *The Princess* is fantastical, metamorphic and redemptive. Critics, with Nature red in tooth and claw uppermost in their minds, have tended to stress the inherent pessimism and darkness of Tennyson's engagement with early evolutionary ideas.[45] This darkness cannot be denied, but Tennyson it seems, like the speakers of *The Princess*, was of 'more than one mind' on the evolution question. Not only does he hold open a range of different political, social and sexual interpretations of the book of Nature in *The Princess* and hold open the conversation that it dramatizes, but the poem itself – comic, hopeful, metamorphic – should be seen as being in open dialogue with its pair, *In Memoriam*. Tennyson's *The Princess* enables us to see how molten evolutionary ideas were in the 1840s, how they were used to support different ideas and ideologies, and how a book like *Vestiges* could shift the ground suddenly towards an evolutionary speculation that was much more progressive, optimistic and forward-looking.

Scientific ideas do not exist in abstract isolation, as a sequence of facts, but in dialogue, in mutual exchange; conversation and debate are where the meaning, the drift, of science is negotiated and made. *The Princess* is a *conversazione* as well as medley; it is a social event. Throughout the poem, evolving conversations between men and women keep discussion of political progress (and avoidance of revolution) bound to ideas of sexual progress and mutability (and affirmation of motherhood). *The Princess* speaks for a radical politics that is both feminist and utopian. Though Ida's political radicalism is revealed to be absolutist and separatist and is modified and softened by the end of the poem, the speculative politics of the poem remain open to the end. The conversation between the Prince and the Princess spills over the end of the poem's supposed closure.

Finally it is this stress that Tennyson places on mixed-sex negotiated exchange in the assimilation of ideas and the drift of science that makes *The Princess* radical in its vision of educational reform. This is Tennyson's final vision: of universities in which men and women not only might have such conversations about science, sexual identity, education and politics in free exchange, but also (and Tennyson of course reveals his ambivalence on this subject too) *must* have them.

Notes

1 Thomas Henry Huxley, *Life and Letters of Thomas Henry Huxley*, ed. Leonard Huxley, 2 vols (London: Macmillan, 1900), 2:338.
2 Hallam Tennyson, *Alfred Lord Tennyson: A Memoir by His Son*, 2 vols (London: Macmillan, 1897), 1:20.
3 Ibid.
4 On the role of conversation in the spread of science in the early part of the nineteenth century see James A. Secord's seminal book: James A. Secord, *Victorian Sensation: The Extraordinary Publication, Reception, and Secret Authorship of 'Vestiges of the Natural History of Creation'* (Chicago: University of Chicago Press, 2000).

5 John Killham, *Tennyson and 'The Princess': Reflections of an Age* (London: Athlone Press, 1958), 235.
6 Ibid.
7 H. Tennyson, *Alfred Lord Tennyson*, 1:44.
8 Killham, *Tennyson and 'The Princess'*, 270.
9 See Adrian Desmond, *The Politics of Evolution: Morphology, Medicine, and Reform in Radical London* (Chicago: University of Chicago Press, 1989).
10 For a more detailed account of Grant and Darwin's informal apprenticeship see Rebecca Stott, *Darwin and the Barnacle* (London: Faber, 2003) and Adrian Desmond, 'Robert E. Grant: The Social Predicament of a Pre-Darwinian Transmutationist', *Journal of the History of Biology* 17, no. 3 (1984): 189–223.
11 Charles Lyell, *Principles of Geology*, 3 vols (London: John Murray, 1830–33), 2:156; a facsimile edition was published by the University of Chicago Press in 1991.
12 H. Tennyson, *Alfred Lord Tennyson*, 1:162.
13 Killham, *Tennyson and 'The Princess'*, 250.
14 For further information on Lyell see Martin Rudwick, *Bursting the Limits of Time: The Reconstruction of Geohistory in the Age of Revolution* (Chicago: University of Chicago University Press, 2005).
15 Robert Chambers, *Vestiges of the Natural History of Creation*, (1844) in *Vestiges of the Natural History of Creation and Other Evolutionary Writings*, ed. James A. Secord (Chicago: Chicago University Press, 1994), 385.
16 A. Tennyson to E. Moxon (15 November 1844), in *The Letters of Alfred Lord Tennyson*, ed. Cecil Y. Lang and Edgar F. Shannon, 3 vols (Oxford: Clarendon Press, 1982–90), 1:230.
17 For a fascinating analysis of science in conversation in this period see James A. Secord, 'How Scientific Conversation Became Shop Talk' in *Science in the Marketplace: Nineteenth-Century Sites and Experiences*, ed. Aileen Fyfe and Bernard Lightman (Chicago: University of Chicago Press, 2009), 23–59.
18 B. Disraeli to S. Disraeli (20 January 1845); M. Disraeli to S. Disraeli (19 January 1845), in *Benjamin Disraeli Letters*, ed. J. A. W. Gunn et al., 8 vols (Toronto: University of Toronto Press, 1982–), 4:154–5, cited in Secord, *Victorian Sensation*, 10.
19 F. Nightingale [to ? P. Nightingale] (February 1845), in Edward Cook, *Life of Florence Nightingale*, 2 vols (London: Macmillan, 1913), 1:37.
20 F. Nightingale to Madame Mohl (July 1847), in ibid., 1:42.
21 Secord, *Victorian Sensation*, 177.
22 Robert Bernard Martin, *Tennyson: The Unquiet Heart* (Oxford: Clarendon Press, 1983), 288.
23 Jane Welsh Carlyle to Helen Welsh, 31 January 1845, *Carlyle Letters* 19:15–17 (I have made slight formatting edits for readability).
24 Norman Page, ed., *Interviews and Recollections* (London: Macmillan, 1983), 9–10.
25 Jane Welsh Carlyle to Thomas Carlyle, Saturday (16 August 1845), *Carlyle Letters* 19:148–51.
26 Benjamin Disraeli, *Tancred or The New Crusade* (London: Peter Davies, [1847] 1927), 112–13.
27 See Stott, *Darwin and the Barnacle*, 240–54.
28 For further information on this eight-year period of Darwin's life see Stott, *Darwin and the Barnacle*.
29 See for instance C. Darwin to J. D. Hooker, Letter 924 (5 or 12 November, 1845), *Darwin Correspondence Project*: http://www.darwinproject.ac.uk (accessed 28 February

2013); also C. Darwin to J. D. Hooker, Letter 1532 (25 September 1853); C. Darwin to J. D. Hooker, Letter 1339 (13 June 1850).

30 Elaine Jordan, *Alfred Tennyson* (Cambridge: Cambridge University Press, 1988), 83.

31 See for instance Isobel Armstrong, 'Tennyson in the 1850s: From Geology to Pathology', in *Tennyson: Seven Essays*, ed. Philip Collins (London: Palgrave Macmillan, 1992), 102–40; Dennis R. Dean, *Tennyson and Geology* (Lincoln: The Tennyson Society, 1985); John Killham, *Tennyson and 'The Princess'*; Milton Millhauser, 'Tennyson's *Princess* and *Vestiges*', *PMLA* 69, no. 1 (1954): 337–43; and Glen Wickens, 'The Two Sides of Early Victorian Science and the Unity of *The Princess*', *Victorian Studies: A Journal of the Humanities, Arts and Sciences* 23 (1980): 369–88.

32 Virginia Zimmerman, *Excavating Victorians* (New York: State University of New York Press, 2009), and Michael Tomko, 'Varieties of Geological Experience: Religion, Body and Spirit in Tennyson's *In Memoriam* and Lyell's *Principles of Geology*', *Victorian Poetry* 42, no. 2 (2004): 113–33.

33 Eve Kosofsky Sedgwick, *Between Men: English Literature and Male Homosocial Desire* (Ithaca and New York: Columbia University Press, 1985).

34 Alan Sinfield, *Tennyson* (London: Wiley-Blackwell, 1986).

35 Virginia Zimmerman argues perceptively that the medley form 'alludes to both geological literature and to the Earth itself'. *Excavating Victorians* (New York: State University of New York Press, 2009), 71.

36 *The Edinburgh Review* 90 (July–October 1849): 388.

37 *Fraser's Magazine* 42 (July–December 1850): 250.

38 Chambers in *Vestiges*, ed. Secord, 197.

39 The first use of the word *dinosaur* was in 1830; the word was coined by British comparative anatomist Richard Owen.

40 William David Shaw, *Tennyson's Style* (Ithaca and London: Cornell University Press, 1976), 117.

41 On the impact of new versions of nature see James Eli Adams, 'Woman Red in Tooth and Claw: Nature and the Feminine in Tennyson and Darwin' in *Victorian Studies* 33, no. 1 (1989): 1, 7–27.

42 A. Tennyson, *In Memoriam*, in *The Poems of Tennyson*, ed. Christopher Ricks (London: Longman, 1969), 912, line 15.

43 Elaine Jordan, *Alfred Tennyson* (Cambridge: Cambridge University Press, 1988), 107.

44 Chambers in *Vestiges*, ed. Secord, 386.

45 One particular example of such an interpretation is the early essay by Dennis R. Dean, 'Through Science to Despair: Geology and the Victorians', *Annals of the New York Academy of Science* 360 (1981): 111–36.

Chapter 3

HISTORY, MATERIALITY AND TYPE IN TENNYSON'S *IN MEMORIAM*

Matthew Rowlinson

This chapter aims to unpack and provide a context for the puzzling amalgam of organicism and historicism to be found in the last lines of Tennyson's *In Memoriam*. In those lines, the poem prophesies the evolution of humankind into a 'crowning race / […] / No longer half-akin to brute' ('Epilogue' 128, 133).[1] With respect to the time of this future race, 'all we thought and loved and did, / And hoped, and suffered, is but seed / Of what in them is flower and fruit' (134–6). The tense shift in line 135, which contemplates the events of the poem and the grief and hope it expresses both in the present and from a standpoint of historical retrospect, dramatizes a split temporality which I will argue both characterizes the poem as a whole and also ultimately comes to define its representation of Tennyson's dead friend Hallam, the elegy's subject. In the poem's final mention of him, Hallam appears with respect to the crowning race to come as 'a noble type / Appearing ere the times were ripe' (138–9). In spite of its reference to ripeness, and the mentions of seed, flower and fruit in the lines immediately preceding, the temporality in which these lines set Hallam is not that of the vegetative cycle but that of history. The source of the phrase 'ere the times were ripe' is *Henry IV, Part 1*, when Worcester warns Hotspur – in vain, as it turns out – to set their plot against the King in motion only 'when time is ripe' (1.3.294).[2] The phrase thus aims to define an historical or political conjuncture; for the revolt to succeed, it must await the preparation of the necessary forces. In Tennyson's figure, Hallam, who resembles Hotspur in living and dying too soon, is represented as appearing out of his proper time, before his required conditions have been met.

I

The term 'type' is itself intrinsically historicist. Like its cognates in German, French and Latin, it derives ultimately from the Greek *tupos*, meaning impression. The root of *tupos* is a verb meaning 'to strike'; the underlying metaphor is thus from numismatics, the *tupos* being literally the impression of a seal in wax or the device hammered onto a coin. This etymology remains clearly visible in both of the term's main senses in current English usage. As part of the lexicon of print, the word 'type' retains its literal reference to a technology of impression – and also a crucial ambiguity about whether type is so called because it produces an impression or because it is produced by one.[3] When it refers to a class or kind within a taxonomical system, as in a blood type or a personality type, the term's reference to the idea of impression is more figural; in this sense, objects of a single type are formed according to a common pattern, of which they metaphorically bear the impression or stamp. In English, though not in other languages, this taxonomical sense of 'type' is recent, dating only from around 1840.

Before turning to nineteenth century senses of 'type', I would like to make two summary observations based on what we have seen so far. The first is that the impression constituting the type is the trace of an historical event. It comes into being in a conjuncture that imposes on a material object a shape or a mark previously alien to it. The type thus exists in a temporal order characterized by rupture rather than by organic growth; it comes into being by analogy with the contingent way metal is made into a coin – it could have been made into something else instead – rather than with the necessity by which an acorn grows into an oak. A second preliminary observation may seem to qualify the type's status as an historical trace, however; this is its characteristic of iterability. As the numismatic metaphor built into the term implies, the type reproduces a pattern that is always in principle subject to further reproduction. This is why the term was able to acquire a taxonomical meaning. Objects of the same type are all formed as iterations of a single pattern, like impressions of a single seal.

On the one hand, then, 'type' denotes the trace or impression of a specific event; on the other, a class of objects. For most of the term's history in English, its meaning has derived from the first of these denotations, which, as we will see, is the one at work in the idea of type that prevails in Christian and especially Protestant exegetical practice. The first two definitions of 'type' in Johnson's *Dictionary* (1756) as 1) An 'emblem [or] mark of something' and 2) 'that by which something future is prefigured' are directly rooted in typological exegesis.[4] In the 1830s Charles Richardson's *New Dictionary of the English Language* defined 'type' as 'A sign or mark (made or formed by *striking*), a form, an image [...] ; a mark, figure, letter,'[5] giving the term's etymology more emphasis than had Johnson, but making no reference to the taxonomical sense then being imported from French and German. No British dictionary seems to have given this sense of the term before 1850, though as Paul Farber notes, Noah Webster did so in 1828 in his *American Dictionary of the English Language*.[6]

A full account of typology as a mode of exegesis is beyond my scope in this essay; even nineteenth-century typology is a vast topic.[7] So we continue in a reductive mode by citing the two passages from St Paul that Erich Auerbach, in a classic article, termed the 'basis' of typology. These are 1 Cor 10:6 and 11 'where the Jews in the desert are termed *typoi hemon* ("figures of ourselves"), and where it is written that "these things befell them as figures (*typicos*)."'[8] In these verses Paul establishes the schema for a mode of figural interpretation that became 'one of the essential elements of the Christian picture of reality, history, and the concrete world in general'.[9] In this schema, 'figural interpretation establishes a connection between two events or persons, the first of which signifies not only itself but also the second, while the second encompasses or fulfills the first. The two poles of the figure are separate in time, but both, being real events [...] , are within time, within the stream of historical life.'[10] The crucial points here are that the type exists in historical time and that it links distinct events as instances of a single pattern.

Ambiguity in the term *tupos*, however, which it shares with the English 'type' and with related terms such as 'stamp' and 'seal', produces a crux in the typological conception of history. Any of these terms can equally denote either something that produces an impression or the impression itself as a product. The relation between a type in the history of the Jews and what from the seventeenth century on was termed its antitype in the Christian dispensation has thus always been open to different interpretations. Do we understand the type as a hammer and the antitype as its impression? Or do we by inversion think of the type as an impression that chronologically precedes its own cause? The prefix 'anti' makes the antitype an inversion of the type, as the idea that one is an impression of the other would imply; but the possibility for confusion is suggested by erroneous uses recorded in the *OED* from the seventeenth and the nineteenth centuries where 'type' refers to the Christian dispensation while the prefiguring event in Jewish history is termed an '*ante*-type'.[11] Or are both type *and* antitype the temporal impressions of a single transcendent pattern that reveals itself by iteration through historical time? This appears to be Auerbach's view when he asserts that typology not only establishes a relation between historical events but also points beyond history:

> Figural prophecy implies the interpretation of one worldly event though another; the first signifies the second, the second fulfills the first. Both remain historical

events; yet both [...] have something provisional or incomplete about them; they point to one another and both point to something in the future [...] which will be the actual, real, and definitive event. [...] Thus history, with all its concrete force, remains forever a figure, cloaked and needful of interpretation.[12]

Typology thus deals above all with a real history in which events appear as letters or forms sequentially imposed on yielding matter. This history, however, can be read in different directions, a fact that in itself implies a split between the time in which it unfolds and the time in which it is interpreted. Whether we accept Auerbach's eschatological reading of the type or not, the split between the time of the typological event and the time when its meaning appears guarantees that it will occur only, in his terms, as 'provisional or incomplete'.

II

In Britain in the 1830s and 1840s the crucial shift in the meaning of 'type' was its adoption into natural history as a term in morphology. This development had little to do with the exegetical sense of the term I have been discussing but was rather the result of new British engagement with scientific developments on the continent, especially in Germany and France. In Germany the terms *typ*, *urtyp*, *urbild* and *bauplan*, among others, were used from the late eighteenth century to designate fundamental structures of organic morphology. In Kant, the existence of morphological types was the ground for what he termed a 'daring venture [...] of reason',[13] the speculation that similarities in structure between different organisms might be taken as evidence of actual kinship between them, though he conceded that no evidence had been found for the transmutation of one species into another. The *Naturphilosphie* of Schelling had as its central premise the idea that morphological types reveal themselves more perfectly in successive cycles of creation. Goethe wrote extensively on the morphology of plants, which he believed was defined by a single underlying pattern, the *urtyp*.[14] This position was shared with respect to vertebrates by Carl Gustav Carus, whose work on the vertebrate archetype was a largely unacknowledged source for Richard Owen in England.[15]

Natural science in early nineteenth-century France was if anything even more concerned with the significance of morphological types than it was in Germany, though without the idea of the type's temporal emergence that featured prominently in much German thought. Though he rarely uses the term 'type', Georges Cuvier's division of the animal kingdom into the four orders (*embranchements*) of vertebrates, invertebrates, mollusks and radiata effectively characterizes each of the great divisions by its typical features. In his last major work, the *Natural History of Fishes*, he proceeds by giving an exhaustive account of a single typical species, the perch, in whose conformation the essential features of fish are said to be summed up. Cuvier's apparently pragmatic adoption of a single species as the type of a genus lays the basis for William Whewell, in his *Philosophy of the Inductive Sciences* (1840), for a natural taxonomy in which classes are fixed not by a definition privileging certain functions or parts, in the Linnaean style, but by an example, which he termed the type: 'A type is an example of any class, for instance a species or genus, which is considered as eminently possessing the characteristics of that class.'[16]

From Whewell's intervention stems the centrality of the term 'type' to British and American debates about taxonomy in the natural sciences during the nineteenth century. Different forms of the type concept provided a major bulwark against the Lamarckian view that the demarcations of species and genera in natural history were arbitrary. For Whewell, as for Cuvier of course, temporal development of the type was excluded; though each accepted the fossil evidence that species had in the past undergone extinction, both viewed the basic categories of animal morphology as immutable and believed that the traits defining them were preserved in the new species created to replace the old.

Nonetheless, the temporal status of the type in its theological sense does exert pressure on its uses in English-language scientific discourse. Charles Lyell, who like Whewell adopts Cuvier's conception of the type species, also uses the term in a chronological sense, referring to the typical species of a particular geological era.[17] The idea that a species is typical because it shows clearly the stamp of a particular time or place indeed becomes commonplace in nineteenth-century science; it plays a major role, for instance, in the polygenetic racial science of the second half of the century.

Even among transmutationists, where given the views of Cuvier, Whewell and Lyell one might not expect to find the concept of type, it is in fact pervasive. Robert Chambers's anonymously published *Vestiges of Creation* (1844) is a particularly clear example of the interaction of the natural-historical and the exegetical concepts of the type. Like Lyell, Chambers could write of a geological era as having its own 'master-form or type'.[18] Chambers's account of the progressive development of higher forms of life during the earth's secular history, moreover, presents the forms of life that have successively evolved as related to one another in a system that is repeated from age to age, so that there is a homology between the relations among the species that make up late-emerging and highly organized genera and the relations among earlier forms. Chambers took over this system, known as quinarianism, from W. S. Macleay and W. J. Swainson; it proposed that every division of the animal kingdom could be arranged into five classes, of which 'the most perfect with respect to the general character of the class' was termed the typical.[19]

Other classes were subtypical or aberrant; the whole quinary structure was repeated 'throughout the whole of the animal, and probably also the vegetable kingdom'. [20] The result, bearing in mind *Vestiges*' central argument regarding the successive appearance of the various orders of organic life, is that each form of life appears as a type – in the sense of a prefiguration – of others that are to follow it. Here is Chambers presenting this position with respect to 'the lowly [...] acrites', or polyp. The acrita, he asserts, 'were the first form of animal life on earth; and they appear like all of those which were to follow in five classes'. In the *Polypi natantes*, the *typical* form of polyp, 'we have a sketch of the *vertebrata*. The acrita thus appear as a prophecy of the higher events of animal development' and 'shew that the nobler orders of being, including man himself, were contemplated from the first'.[21] When humankind finally appears on the scene, as the typical species of the Mammalia, it is as 'the type of all types of the animal kingdom'.[22]

III

The fusion of the senses of 'type' as taxon and as prophecy to be found in Chambers is most familiar today in a work strongly influenced by *Vestiges*, Alfred Tennyson's *In Memoriam*, published in 1850. Tennyson's poem is crucially concerned with the fossil evidence for the extinction of species, which it terms 'types', and it ultimately adopts Chambers's belief that extinction is a necessary part of the evolution of lower forms of life into higher ones. It also incorporates references to biblical types, like the image of water rising in the rock that organizes section 131. Its most systematic use of typology, however, is not as a source of images but as a formal principle; as the poem unfolds in time, it establishes multiple patterns of recurring events and motifs, each of whose appearances reinterprets all of the others. The interpretation of the meanings that these repetitions bring into being is at one level the poem's main action.

Tennyson's poem thus has a typological structure; as we saw earlier, one of the effects of such a structure is to split or multiply temporalities. Though the type is constituted by time, it exists in more than one present and belongs to temporal sequences that can run in different directions. This is why *In Memoriam* is structured by multiple and asynchronous internal calendars. As is well known, the poem presents a process of mourning that extends over three years, with the passage of time marked by the turning of the seasons and by the recurrences of Christmas and the anniversaries of Hallam's birth and death. This calendar compresses the actual period of Tennyson's work on the poem, which ran from the year of Hallam's death in 1833 till shortly before its publication in June of 1850. This second calendar, however, does not simply remain outside the poem; rather, it is incorporated into it, establishing a second principle of organization that operates alongside the temporal markers I have just mentioned. At the poem's initial publication in a stand-alone volume, the title page naturally bore the year of publication. After this it opened with an untitled section – which A. C. Bradley later made it conventional to term the 'Prologue' – that makes a retraction of the poem to follow, referring to it as the 'confusions of a wasted youth' (42). This section, uniquely in the poem, bears a date, 1849, in Arabic numerals, marking its retrospective view of what follows. In the authorized editions of the poem – on a separate page in the stand-alone editions – there then follows a dedication: 'In Memoriam A. H. H.', with the year of Hallam's death in Roman numerals.[23] From this second beginning the poem then proceeds in both directions: forward along its three year calendar and also back to the dated retraction we have already read, and ultimately back to the title page, where the dedicatory act performed by the formula 'In Memoriam' is cited to become the poem's title.

In Memoriam is by no means the only mid-Victorian text with conflicting chronologies. The trait may appear wherever we find explicit or implicit reference to the forms of biblical narrative. Dickens's *Bleak House*, for instance, features a typological structure in which the old Bleak House prefigures and is superseded by the new; this apparently simple temporal logic is, however, complicated by the novel's two narrators, who use different tenses, and by the way the narrative is shot through with hints of anachronism, like the fantasy of a 'megalosaurus' on Holborn Hill in the opening paragraph,[24] and recurrent allusions to a coming apocalypse in a world where nothing ever seems to change.

The conjunction of different temporal systems in *In Memoriam* is the topic of a section where the poem reflects explicitly on its own structure and meaning. Passages like this one, where the poem appears as its own interpreter, necessarily have the character I have been discussing, of belonging to more than one temporal schema, since they occupy a position at once within the poem's diegesis and outside it. In section 121, the conjunction in a single textual moment of multiple temporal systems is the topic of the poem's auto-exegesis. The section's major trope is apostrophe, the figure of presence *par excellence*, in which the direct address of speaker to auditor presumes their coexistence in a single place and time. The addressees in this poem are themselves in part personified moments in time, in part celestial bodies – Hesper and Phosphor, the morning and evening stars, each of which is represented as watching and listening to the characteristic events of the hour at which it appears. Nonetheless, their character as personified moments in time notwithstanding, both Hesper and Phosphor, morning and evening, have a temporal existence that is principally defined as the mediation of another moment, of the full presence or absence of the sun in the coming day or night.[25] In this sense both Hesper and Phosphor are types – the point is clearest in the address to Phosphor: 'Behind thee comes the greater light' (12). Hesper and Phosphor are thus equally lights whose meaning is given by their relation to the greater light that either follows or precedes them, as antitype to type.

More fundamentally, the typological ordering of section 121 appears in its mythological and linguistic syncretism. Typology has been since St Paul the dominant Christian approach to Hebrew scripture. It is necessarily an exegetical practice of translation and refiguration in which putatively erroneous or partial texts are read as anticipating their fulfillment or completion by true ones. The difference between texts in a typological scheme is always, as we have seen, understood as a difference between historical moments. If section 121 begins by personifying the morning and evening stars, its eventual argument depends upon the knowledge that these personifications are merely different names and figures for a single celestial body: 'Sweet Hesper–Phosphor, double name / For what is one' (17–18). The argument also requires us to recognize that Venus, the modern name of this body, was previously the Latin name of the goddess of love. Behind the Greek names of the personified morning and evening, there appears the name in a different language of one of the archetypal mourners of the elegiac tradition, whose grief over the dead Adonis is an explicit *topos* in the genre from its beginning in Moschus's 'Lament for Bion' up to Shelley's 'Adonais' and is an implicit point of reference for the series of female mourners in Tennyson's own poem.[26] Finally, behind Venus as goddess and mourner, there appears Venus as a planet, whose shining by the reflected light of the absent sun makes it a figure for the elegist himself.

The central figure of *In Memoriam* section 121 links past and present as repeated appearances of the same thing, using a typological schema to assert the continuity of the poet's identity and, therefore, the formal coherence of the poem by which that identity is represented. If its primary concern is with coherence within the temporal span of a single human life, though, the typological schema the poem elaborates nonetheless also encompasses a history whose phases – we might term them animist, theological and scientific – are made to appear as types of one another and ultimately of the poet's life in the present.

Our reading, though, shows this present's indeterminacy. We began with the claims that section 121 has apostrophe as its organizing trope and that this trope guarantees and indeed constitutes for the poem a series of presents and presences. The opening two stanzas are set at evening and addressed to Hesper as evening's personification; the next two are set at morning and addressed to Phosphor. What time, however, is the setting of the final stanza, addressed to 'Sweet Hesper–Phosphor' (17)? Here the poem reveals the succession of present moments apparently constituted by apostrophe as a fiction and sets itself in a temporal order of a different, apocalyptic, kind where present and past coexist. This consequence of the type's temporal schema extends beyond the borders of this single lyric to disturb or render indeterminate the entire diegesis of Tennyson's lyric sequence. When this section asserts the unity of the speaker's past and present, it is radically unclear what past and what present it refers to. Is the poem's present a morning of joy that has succeeded the evening of mourning in which the poem's opening sections are set? Or is the section's present defined by its opening stanzas, which would then appear in contrast to the joyous dawn of the poet's life with Hallam? In light of its double address neither reading of the section will suffice, and we are led to view it as locating within the poem a position from which its structuring narrative can be seen from outside. An apocalyptic view – indicated as such by the allusion to Revelation 22:16 in the identification of Hesper–Phosphor as the double name 'for what is one, the first, the last' (18) – thus emerges as the only one in which the type can appear as fulfilled or complete.

IV

In Memoriam closes – if we ignore the opening, also a kind of close – by anticipating a 'coming race' ('Epilogue', 128) that is to close the evolutionary sequence. The passage echoes Chambers, who concludes his chapter on 'Animated Nature' by speculating that 'our race' might be 'but the initial of the grand crowning type' (276). Chambers's odd figure, with its clutter of terms of art from printing, at once imagines the coming race as closing an evolutionary narrative and as the final character in a passage of print – that is to say it represents 'our race' and the one to come as existing both in temporal succession and simultaneously. Though humanity may be best adapted to the present state of things, Chambers writes, the external world will in time undergo change, whereupon 'there may be occasion for a nobler type of humanity, which shall complete the zoological circle on this planet' (276). The figure of the circle here, which anticipates the coincidence of beginning and ending in the unnumbered first section of Tennyson's poem, again leaves it equivocal whether Chambers is referring to the completion of a temporal narrative or of a taxonomical schema.

In the final stanzas of Tennyson's poem, he echoes Chambers's speculations and makes a last reference to Hallam, remembering him with reference to the form of humanity that is to come as 'a noble type / Appearing ere the times were ripe' ('Epilogue', 139–40). In taking over Chambers's narrative, Tennyson shifts the term 'type', as we have seen is always possible, from the future to the past and supplements its natural-historical sense with its earlier reference to prefiguration. He makes Hallam, as it were, the type of a type; appearing before the proper moment in Chambers's 'zoological circle', he can only prefigure the class of which he is an example. With this ending the poem revises a figure that goes back to the beginnings of elegy, the association of the lost friend with the cycle of plant life, so that in death he can be represented as a flower cut down too soon. In Tennyson's final version of the figure, though, it is not Hallam's death but his birth, and indeed his entire earthly existence, that happen too soon. Moreover, in spite of the references to ripeness, as well as to seed, flower and fruit, in the lines immediately preceding, the temporality in which the epilogue sets Hallam is not that of the vegetative cycle at all but that of history. Though Tennyson's usage refers to both the morphological and the theological senses of 'type', neither sense allows for the idea of the type as premature, which transforms the idealist and theological ideas of history that the concept brings with it into something more ironic.

As we have seen, *In Memoriam* is structured by multiple internal time-schemes. As a result, with respect to the different calendars that the poem incorporates, none of its events occur at a proper time. This characteristic prefigures the view of Hallam's life that it reaches at its close. It also points to a general feature of Tennyson's historical poetics. His works are often striking in their topicality or in some way endowed with what we could call a dateline. In its representation of domestic life, and in its political references – updated before publication to refer to the events of 1848[27] – *In Memoriam* is much more directly a poem of its time than, say, 'Adonais', 'Thyrsis,' or 'Ave atque Vale', three other hypercanonical elegies of the nineteenth century. Rather than linking Tennyson's poems to a single historical date of origin, however, their topical specificity more characteristically produces in them a kind of internal difference or anachronism. Much has been written on the problem of actually assigning a date to many of Tennyson's poems, which were often composed over decades. Contradictory or ambiguous temporal positions are moreover a recurrent formal and thematic preoccupation in his work. In major dramatic monologues, what we call the dramatic situation can often also be read as the rehearsal or repetition of a situation, a problem that divides the very notion of a 'situation' against itself. I am summarizing here claims I've made elsewhere with respect to 'Ulysses' and 'Tithonus';[28] these two monologues also offer particularly elaborate instances of a trope that preoccupies Tennyson throughout his career, that of the double star Hesper–Phosphor and the chiastic interchange of morning and evening, beginning and ending.

Tennyson's poems can thus have a contradictory relation to any specification of a time or date. One manifestation of his interest in the temporal location of his poems is the prominence among them of anniversary observances; these include official poems such as 'On the Jubilee of Queen Victoria', late poems commemorating the 600th anniversary of Dante's birth and the 1900th of Virgil's death, and also birthday epistles to friends and contemporaries, such as those to Edward FitzGerald and W. G. Palgrave. The addressees of both of these epistles died before receiving them; the poem to FitzGerald moreover got his age wrong, complimenting him on a 75th birthday he never in fact saw. Tennyson's response to FitzGerald's death was to add a pendant to his poem, transforming it into an elegy and incorporating into it a figure for its own belatedness: their addressee's death has 'made the rhymes, / That missed his living welcome, seem / Like would-be guests an hour too late, / Who down the highway moving on / With easy laughter find the gate / Is bolted, and the master gone' (67–72).[29] Without being reductive, it is worth saying that Tennyson could have suppressed the original poem to FitzGerald; he could also have published it without modification and allowed the fact of publication to itself to sever it from its original context and addressee. In print, the address of a birthday or other epistle becomes to a greater or lesser extent a fiction; as a fiction the poem survives and can indeed be said to presuppose the death of the figure whose birth it putatively commemorates. For Tennyson, however, the poem's connection to its original occasion is sufficiently strong – the poem is so topical – that the forestalling of its address to FitzGerald transforms it not into a fiction but into a dead letter. The double temporality of a poem that is at once inseparable from its occasion and too late for it gives a paradigm that has a broad application in Tennyson's work, especially *In Memoriam*.

Poems of direct lyrical address and poems of anniversary commemoration indeed play important roles in *In Memoriam*, functioning as structuring devices in the sequence as a whole. *In Memoriam* comprises 131 numbered lyrics, framed by an unnumbered proem and an epilogue. The great majority of these lyrics are in the present tense; their status as utterances having a specific temporal location is in many cases emphasized by Tennyson's use of the figure of address to endow a section with a specific dramatic situation: 'Dark house, by which once more I stand [...] '. This line, the opening of section 7, reminds us how frequently, moreover, the situations in *In Memoriam* repeat an earlier situation or commemorate a past event. Tennyson stands before Hallam's house as he often stood while Hallam was alive; later in the sequence, in section 119, he will return again to the same place, initially to address the house once more but ultimately to address Hallam himself, who in the earlier section appears only as an absence: 'He is not here' (9).

This kind of repetition-with-a-difference is the fundamental principle by which *In Memoriam* links its component sections, with their discrete lyric presents, into a narrative. Every event in the poem is thus a kind of commemoration, so it is not surprising that prominent among them are several explicit commemorations of anniversaries. As we know, the sequence dramatizes a process of mourning that extends over three years; *In Memoriam*'s sheer length entails a significant revision of elegiac convention. In a departure both from the genre's Classical models and from Milton's Christianized version, neither the natural nor the sacred calendar is found in Tennyson's poem to be a source of consolation. The poem marks the observation of three Christmases, but without the Christian promise of personal immortality. Nor does it find any promise of individual rebirth in the natural world; by its structure and its length, it affirms that the natural cycle leading from death to life also leads back to death again. The poem's calendar measures the homogeneous empty time of bourgeois history; the meanings of its most important dates, like the anniversaries of Hallam's birth and death, are contingent rather than motivated by theology or cosmology. As Tennyson notes in section 99, the date of Hallam's death is for some people a birthday or a wedding anniversary. Even the three sections set on successive Christmases represent it as a secular family gathering rather than a religious observance. The sacred calendar, like the solar one, functions as an apparently neutral ground that does not determine the events the poem superimposes on it.

The poem's contents are thus arrayed along several distinct and incompatible temporal axes. By dating the dedication and the proem, Tennyson allows the retraction performed by the latter to encompass the whole chronological scheme organizing the material that follows, which it proleptically exposes as a poetic fiction. Like the anniversary poem for FitzGerald, individual sections of *In Memoriam* are occasional poems whose occasions don't take place. In this respect the whole sequence remains in the traumatic situation described in section 6, where the poet recalls writing before the news has come of Hallam's death and compares his work, 'wrought / At that last hour to please him well' (17–18), to the self-adornment of a young girl dressing to meet her lover – all unknowing that 'the curse / Had fallen, and her future Lord / Was drowned in passing through the ford, / Or killed in falling from his horse' (37–40). The poem itself participates in the illusion it exposes, referring to the girl's 'future Lord' even as it makes clear that the future in question no longer exists. Though Tennyson writes here in the past tense of his preparations for the reunion with Hallam that never came, the poem nonetheless still occupies a present structured by a missed occasion.

In the middle decades of the nineteenth century, the term 'type' designated in one field the imprint *on* history of an ideal form and in another the working out *in* history of a providential design. Tennyson's type, like his poetry, *has* a proper time with respect to both of these concepts of history – but *misses* it.

With respect to these concepts it designates something lost, a conjuncture that they foreclose. Given the source of the phrase 'ere the times were ripe' in *Henry IV, Part 1*, we can see in this foreclosure not only the trace of a traumatizing personal loss but also the possibility of an oppositional or even revolutionary relation to history.[30]

Notes

1. Except where noted, citations of Tennyson's poetry are drawn from Christopher B. Ricks, ed., *The Poems of Tennyson*, 2nd ed., 3 vols, Longman Annotated English Poets (Harlow: Longman, 1987).
2. G. Blakemore Evans et al., eds, *The Riverside Shakespeare* (Boston: Houghton Mifflin, 1974). After the mid-nineteenth century, Shakespeare's phrase became a cliché, and it is now often used without any awareness of its source. Prior to 1850, however, I have not found any use of the phrase that does not make explicit reference to *Henry IV, Part 1*. The Shakespeare allusion can thus be assumed to be more active in Tennyson's lines than it appears now.
3. According to the *Oxford English Dictionary*, the word 'type' was incorporated into the lexicon of print only in the early eighteenth century. From the beginning of the print era, pieces of movable type were cast in matrices produced by striking with a punch. The punch was the original engraved form of which every piece of type was the impression. For a description of type-making see Philip Gaskell, *A New Introduction to Bibliography* (New York: Oxford University Press, 1972), 10–11.
4. Samuel Johnson, *A Dictionary of the English Language*, 2nd ed., 2 vols, (London: printed by W. Strahan, for J. and P. Knapton; T. and T. Longman; C. Hitch and L. Hawes; A. Millar; and R. and J. Dodsley, 1755–56), 2:962.
5. Charles Richardson, *A New Dictionary of the English Language* (London: William Pickering, 1839), 825 (emphasis in the original).
6. Paul Lawrence Farber, 'The Type-Concept in Zoology in the First Half of the Nineteenth Century', *Journal of the History of Biology* 9, no. 1 (1976): 93.
7. The two major treatments of Victorian typology are Herbert L. Sussman, *Fact into Figure: Typology in Carlyle, Ruskin, and the Pre-Raphaelite Brotherhood* (Columbus: Ohio State University Press, 1979) and George P. Landow, *Victorian Types, Victorian Shadows: Biblical Typology in Victorian Literature, Art, and Thought* (Boston: Routledge and Kegan Paul, 1980). For an important discussion of language, natural history and the type in *In Memoriam*, see Isobel Armstrong, *Victorian Poetry: Poetry, Poetics, and Politics* (London: Routledge, 1993), 247–63.
8. Erich Auerbach, 'Figura', in *Scenes from the Drama of European Literature: Six Essays* (1959; repr., Minneapolis: University of Minnesota Press, 1984), 49.
9. Ibid., 33.
10. Ibid., 53.
11. I owe this reference to Peter Stallybrass.
12. Auerbach, 'Figura', 58.
13. Immanuel Kant, *The Critique of Judgement*, trans. James Creed Meredith, 2 vols (Oxford: Clarendon Press, 1952), 2:79.
14. On Goethe, and on plant morphology in German thought of the Romantic era more generally, see Robert J. Richards, *The Romantic Conception of Life: Science and Philosophy in the Age of Goethe* (Chicago: University of Chicago Press, 2002).

15 On Carus and Owen, see Nicholas A. Rupke, 'Richard Owen's Vertebrate Archetype', *Isis* 84 (1993).
16 William Whewell, *The Philosophy of the Inductive Sciences, Founded Upon Their History*, 2nd ed., 2 vols (London: John W. Parker, 1847); 1:494.
17 Charles Lyell, *Principles of Geology, Being an Attempt to Explain the Former Changes of the Earth's Surface by Reference to Causes Now in Operation*, 3 vols (London: John Murray, 1830–33), 3:50.
18 Robert Chambers, *Vestiges of the Natural History of Creation* (1844; repr., Leicester: Leicester University Press, 1969), 84.
19 Ibid., 240.
20 Ibid., 242.
21 Ibid., 249–50 (emphasis in the original).
22 Ibid., 272.
23 For this dedicatory page, which Ricks's edition of Tennyson does not reproduce, and for other details of the poem's publication history, see Alfred Tennyson, *In Memoriam*, ed. Susan Shatto and Marion Shaw (Oxford: Clarendon Press, 1982).
24 Charles Dickens, *Bleak House*, ed. Stephen Gill (Oxford: Oxford University Press, 1996), 11.
25 For a study of mediation in the English 'Hesperian' poem, see Geoffrey Hartman, 'Poem and Ideology: A Study of Keats' s "To Autumn"', in *The Fate of Reading and Other Essays* (Chicago: University of Chicago Press, 1975).
26 On the female mourner in male elegy, see Peter M. Sacks, *The English Elegy: Studies in the Genre from Spenser to Yeats* (Baltimore: Johns Hopkins Press, 1985). Sacks's chapter in this book on *In Memoriam* is the fullest treatment of the poem's typological structure.
27 After the trial edition, Tennyson revised section 127 so that it referred to 'The red fool-fury of the Seine' (7) as coming 'thrice again' rather than 'once' – presumably recalling the three revolutionary years of 1789, 1830 and 1848.
28 Matthew Rowlinson, *Tennyson's Fixations: Psychoanalysis and the Topics of the Early Poetry* (Charlottesville: University of Virginia Press, 1994).
29 The poem frames 'Tiresias', which is also about words as dead letters.
30 The cardinal event of Tennyson's friendship with Hallam was their trip to the Pyrenees in the summer of 1830 to bring money and dispatches in support of an abortive rising against the absolutist monarchy of Ferdinand VII. Recollections of the landscape through which they travelled recur throughout Tennyson's subsequent poetry, most vividly in 'In the Valley of Cauteretz', written on the occasion of a return to the Pyrenees in 1861. This poem explicitly memorializes the original trip with Hallam, whose date it mistakes. For details of this trip and of the disastrous outcome of the movement it supported, see A. J. Sambrook, 'Cambridge Apostles at a Spanish Tragedy', *English Miscellany* 16 (1965). On revolutionary bad timing, see Slavoj Žižek, *The Sublime Object of Ideology* (London: Verso, 1989), 58–64.

Chapter 4

DARWIN, TENNYSON AND THE WRITING OF 'THE HOLY GRAIL'

Valerie Purton

In 1888 Algernon Charles Swinburne published in the *Nineteenth Century* a short spoof in the style of the 'Did Bacon Write Shakespeare?' articles popular at the time, in which he claimed energetically that the real author of Tennyson's poems was actually Charles Darwin.[1] He cites as proof 'the well-known passage from *Maud* beginning with what we may call the pre-Darwinian line – "A monstrous eft was of old the lord and master of the earth" as well as "the celebrated lines which describe Nature as "so careful of the type, so careless of the single life" (129).' Earlier chapters of the present volume, in their explorations of Tennyson's response to pre-Darwinian evolutionary debates, especially in his reading of Charles Lyell and of Robert Chambers, have revealed that there was a genuine perceptiveness behind Swinburne's comedy. The present chapter focuses on a direct encounter between Tennyson and Darwin and will argue (giving unexpected, if extremely limited, support to Swinburne!) that there may be a link between the only documented meeting of these two iconic figures, on 17 August 1868, and Tennyson's subsequent completion of 'The Holy Grail' idyll in what Emily Tennyson described as 'a breath of inspiration' during the following three weeks.

The Holy Grail episode had been on Tennyson's mind for over a decade: to him it was the key to the whole cycle of *Idylls of the King*, but he demurred year after year, doubting, he said, 'whether such a subject could be handled in these days, without incurring a charge of irreverence'.[2] This chapter will explore the state of the religion and science debate in the 1860s in an attempt to explain Tennyson's difficulties; it will then turn to Tennyson's reading 'with intense interest'[3] of *The Origin of Species*, in November 1859, and will produce a re-reading of 'The Holy Grail' through the prism of the *Origin*. My argument is that the encounter with Darwin may have helped give Tennyson the confidence to explore a more sceptical reading of the grail theme, while reassuring him that this need not be a revolutionary act.

Tennyson's own generation certainly thought of him as 'the Poet of Science': Huxley's famous encomium, that Tennyson was 'the first poet since Lucretius who understood the drift of science'[4] sums up popular opinion at the poet's death.[5] The obituary in the scientific journal *Nature* praised him as 'the Poet who, above all others who have ever lived, combined the love and knowledge of Nature with the unceasing study of the causes of things and Nature's laws'. His achievement, the obituarist declared, was to show that 'science and poetry, far from being antagonistic, must forever advance side by side.'[6] The astronomer Norman Lockyer, author of that unsigned obituary and a friend of Tennyson, summed up his contemporaries' assessment of the poet when he published, together with his daughter Winifred, a volume of extracts from Tennyson's poetry entitled *Tennyson as a Student and Poet of Nature* – 'nature' in this context meaning something closer to 'natural science' than to pastoral poetry.[7] More recent critics have found a much more defensive Tennyson. A. J. Meadows ends an essay on 'Tennyson and Science' in the *Notes and Records of the Royal Society* (1992) with the comment that 'the closer acquaintance that Tennyson had with science, the more worried he seems to have become about its implications.'[8] He concludes that the 'growth of science and the increasing confidence of scientists during his lifetime gave [Tennyson] increasingly to wonder how science and poetry, as he practised it, were to be reconciled.'[9] 'Poetry as he practised it' presumably means poetry reliant for its effect on conveying a sense of the mystery of the world and of belief in human immortality. Meadows signally misses the importance of 'Honest Doubt' as an impulse behind some of the strongest writing – not only of Tennyson but of his contemporaries.

The years just before and after the publication of *The Origin of Species* found many of the great mid-Victorian writers moving towards scepticism. John Ruskin (discussed by Clive Wilmer in Chapter 8) had completed the last volume of *Modern Painters* in 1858. However, that volume already contained what Kenneth Clark sees as a bold declaration of humanism: 'Therefore it is that all the power of nature depends on the subjection of the human soul. Man is the sun of the world; more than the real sun.'[10] Post-Darwinian novels began almost immediately to evince what Beer and Levine have identified as 'Darwin's plots' – beginning with *Great Expectations* (1860–61) where those embodiments of intractable virtue, Oliver Twist and Little Nell, give way to all too malleable figures who are shown adapting to circumstances in order to survive. 'What could I become with these surroundings?' meditates Pip, on the warping influence of his time at Miss Havisham's.[11] The novel genre itself responded to the mood of the age. Ronald Thomas, George Levine and others have pointed to the way in which the new ideas of Darwinian biology contributed to the sudden emergence in the 1860s of a new subgenre, the sensation novel,[12] elements of which are already evident in *Great Expectations* in 1861–2. The rise of the sensation novel is of particular relevance to any discussion of Tennyson, 'steeped' as he confessed himself to be 'in Miss Braddon'.[13] Mary Elizabeth Braddon, whose own life seems a potent case study for the operation of Darwinian laws, published *Lady Audley's Secret* in 1862. This tale of the Victorian heroine turned murderess, who lies and manipulates in order to survive, seems to have horrified and fascinated Tennyson and perhaps even contributed to that visceral fear of degeneration evident from Allingham's diary in 1867, where he records Tennyson's diatribe against 'Women in towns, dangers to health, horrible diseases, T[ennyson] would have a strict Contagious Diseases Act in force everywhere.'[14] Fear of degeneration was later in the century to become another aspect of Darwin's legacy. Interestingly, it is in the same entry that Allingham also records Tennyson's desperate search for convincing proof of human immortality:

> He spoke of immortality and virtue [...] 'Sometimes I have a kind of hope.' His anxiety has always been great to get some real insight into the nature and prospects of the Human Race. He asks every person that seems in the least likely to help him in this, reads every book. When *Vestiges of Creation* appeared [1844] he gathered from the talk about it that it came nearer an explanation than anything before it. T. got the volume and (he said to me), 'I trembled as I cut the leaves. But, alas, neither was satisfaction there.'[15]

The story of Tennyson's move from reading Charles Lyell's *Principles of Geology* in despair to reading Robert Chambers's *Vestiges of Creation* and recapturing hope is powerfully told by Eleanor Mattes – but the linearity of the Mattes narrative itself seems limiting, suggesting a very pre-Darwinian emotional trajectory.[16] Even if Tennyson's pre-Darwin response to science and specifically to the evolutionary debate could be figured in such a simple way, his reaction to the publication of *The Origin of Species* reverberated through the following decades to produce a bewildering range of poetry.

Tennyson's interaction with the scientists of his age was considerable. Sir John Herschel was a friend, as was Professor Richard Owen, doughty adversary of Darwin. Owen himself loved Dickens's novels as well as Tennyson's poetry, and Emily Tennyson's record of his visit to Farringford includes a very Dickensian anecdote of Owen's (an egregious example of the interaction of literature and science):

> July 23rd [1865]
> Farringford. Professor Owen arrived. A[lfred] went with him to Brightstone. They spread out their luncheon on Mr Foxe's lawn and looked at the great

dragon (a Saurian reptile dug up at Brooke) which was new to the Professor, and which quite answered his expectations. He never saw one so sheathed in armour, and thought of calling it Euacanthus Vectanius. Most interesting he was. The story of his medical student days. Of the negro's head which he had been carrying slipping from under his arm, bounding down the hill and bursting through a window into the midst of a quiet family at tea: their horror: his rushing in after the head without a word, and clutching at it and 'bolting', was very ghastly.[17]

The Tennysons had been involved early in the rising tide of Honest Doubt. It was in 1853 that F. D. Maurice's *Theological Essays* had criticized the doctrine which later was to cause Darwin such agony – that the spiritually unreceptive would suffer eternal damnation. ('A damnable doctrine', Darwin was to call it.[18]) Tennyson had known Maurice in Cambridge as, with John Sterling, the doyen of the early Cambridge Apostles. He had read Maurice's works as they appeared, invited him to his son Hallam's christening and maintained his usual loyalty to his friends. When Maurice lost his Chair at King's College London as a result of the *Theological Essays*, Tennyson addressed him in a sonnet, written in 1854, inviting him to Farringford – 'To the Rev. F. D. Maurice'.[19] The Maurices stayed at Farringford in 1858. Charles Kingsley responded to both Darwin and Maurice. He wrote to Maurice that he 'was utterly astonished at finding in page after page things which I had thought, and hardly dared confess to myself, much less to preach'.[20] Tennyson and Darwin both struggled with the notion of eternal damnation.[21] Other writers had different concerns – and the climate of the age quickly seems to have altered to allow freer expression. Herbert Spencer in *Principles of Biology* (1864) boldly extended Darwin's ideas to the higher animals. The theologians who contributed to the liberal *Essays and Reviews* (1860) did not suffer Maurice's fate – indeed, one of them went on to become Archbishop of Canterbury. George Eliot in *The Mill on the Floss* (1860) moved uneasily between scientific realism and myth. Her conclusion, after the drowning of the brother and sister, Maggie and Tom, in the great flood on the River Floss, is bravely Darwinian and unillusioned:

> Nature repairs her ravages, but not all. The uptorn trees are not rooted again; the parted hills are left scarred: if there is a new growth, the trees are not the same as the old, and the hills underneath their green vesture bear the marks of the past rending. To the eyes that have dwelt on the past there is no thorough repair.[22]

***

On the Origin of Species shared the bookshop windows in 1859 with the first major collection of Tennyson's *Idylls of the King*, containing 'Vivien', 'Elaine', 'Enid' and 'Guinevere'. Both works were the result of long gestation and were, for a variety of reasons, long delayed. They were both earnests of larger works to come, though in the event only Tennyson was to complete that larger work. Tennyson's early *Idylls* are absolutist: the subtitle, 'The False and the True', reveals the simple binary morality underpinning the poems. By the time he came to publish the second batch of *Idylls*, in 1868, Tennyson had read Darwin. If post-Darwinian novels contain 'Darwin's plots', then it seems likely that evidence of Darwin's influence will also be present in post-Darwinian poetry. Certainly, these later *Idylls*, 'The Coming of Arthur', 'The Passing of Arthur', 'Pelleas and Ettarre' and 'The Holy Grail', display quite a different, darker world view. As Graham Hough has pointed out, 'Darwin was concerned with the mechanism, Tennyson with the moral and theological consequences: but they are, quite independently, following the same movement of thought.'[23] The structure of the *Idylls* as Tennyson worked on it during the later decades of his life is obviously affected by his reading and sometimes misreading of Darwin and by his struggling with the concept of evolution. The narrative of these later *Idylls* is not linear and progressive but rather concerned with transience. The fall of Camelot is already implicit in its rising: 'The Coming of Arthur' begins with a vision of impermanence, of endless change rather than fixity, which may well owe something to Tennyson's reading of Darwin:

> For many a petty king ere Arthur came
> Ruled in this isle, and ever waging war
> Each upon other, wasted all the land;
> And still from time to time the heathen host
> Swarmed overseas, and harried what was left.
> And so there grew great tracts of wilderness,
> Wherein the beast was ever more and more,
> But man was less and less, till Arthur came.
> For first Aurelius lived and fought and died,
> And after him King Uther fought and died,
> But either failed to make the kingdom one.
> And after these King Arthur, *for a space*,
> And through the puissance of his Table Round,
> Drew all the petty princedoms under him,
> Their king and head, and made a realm, and reigned.[24]

'For a space' is a key phrase. As a young man, Tennyson was already fascinated by the processes of change. He finds this in Heraclitus, whom he commemorates in an early poem written even before he had read Charles Lyell, and published in 1830:

> All thoughts, all creeds, all dreams, are true;
> All visions wild and strange.
> Man is the measure of all things
> Unto himself. All truth is change.
>
> All men do walk in sleep, and all
> Have faith in that they dream:
> For all things are as they seem to all,
> And all things flow like a stream.
>
> There is no rest, no calm, no pause,
> Nor good nor ill, nor light nor shade,
> Nor essence nor eternal laws:
> For nothing is, but all is made.
>
> But if I dream that all these are,
> They are to me for that I dream:
> For all things are as they seem to all,
> And all things flow like a stream.[25]

Tennyson, on this evidence, had little difficulty in imaginatively grasping that aspect of the *Origin* with which his contemporaries struggled most: the 'permanence of impermanence' implied in the struggle for existence. He has no prior vision of liberal progressivism, of George Eliot's meliorism, to give up, and like Darwin himself, he sees no reason to despair at the thought of endless fluidity. In 'The Coming of Arthur' there is despair in the midst of hope. The corollary, though, is that in 'The Passing of Arthur' there is hope in the midst of despair. After 'that last weird battle in the West'[26] comes one of the most brilliant passages in Tennyson, anticipating both the First World War paintings of Paul Nash and the poetry of T. S. Eliot. What it implies is that, in a world ruled by evolutionary principles, even despair is impermanent:

> Only the wan wave
> Brake in on dead faces, to and fro
> Swaying the helpless hands, and up and down
> Tumbling the hollow helmets of the fallen,
> And shivered brands *that once had fought with Rome*,
> And rolling far along the gloomy shores
> The voice of days of old and days to be.[27]

The *Idylls of the King* needs to be set in its own time – or rather, in *their* own times because, of course, these twelve sprawling narrative poems were written at various stages of Tennyson's life, between 1833 and 1885. They rework the myth of King Arthur in ways which are peculiarly Victorian, and as such they are a treasure-house of insights into attitudes to women (Guinevere's adultery with Lancelot being presented as *the* reason for the collapse of empire), attitudes to empire itself (amazingly unillusioned and indeed doom-laden), and (of particular relevance to this chapter) attitudes to religion and science. Because of their mode of production, the *Idylls* raise a fascinating theoretical question, central to the religion–science debate: What exactly is narrative? What counts as a beginning, a middle and an end? Is there such a thing as teleology? The *Idylls* begin at the end, in 1833, with the 'Morte D'Arthur' ('So all day long the noise of battle rose'), Arthur's last battle against Mordred, ending in his death – or rather, his being carried off to Avilion in a barge 'Where I will heal me of my grievous wound'.[28] This is the mythical encoding of a very intense private elegy to Arthur Hallam, who had died so suddenly on 15 September 1833. The *Idylls* end in the middle, in 1885, with 'Balin and Balan', also to do with intense grief at a death, as the two brothers, unknowingly, kill each other in battle. I have argued elsewhere that this too is an intimate personal elegy, this time to Tennyson's beloved brother Charles.[29] Though not published until 1885, the poem turns out to have been written much earlier, close to 1879, when Charles died. The whole cycle, then, debates whether individual human achievement has any meaning. The *Idylls* can be read either synchronically (as an 'eternal whole' already present in the young Tennyson's mind and simply waiting to be transcribed) or diachronically (as changing through time, bearing strong textual traces of contemporary debates) and as adapted to contingent circumstances. The question of how to read the *Idylls*, then, engages with the debate central to both Darwin and Tennyson: should life itself be read teleologically, as possessing a shape and meaning of its own, or contingently, as a series of adaptations to changing circumstances?

'The Holy Grail' was, said Tennyson, 'the key to all the *Idylls*'.[30] For decades, however, he couldn't bring himself to approach the subject. Hallam Tennyson records from his mother's journal that his father 'had the subject in his mind for years, ever since he began to write about Arthur and his knights'.[31] He was hampered, Hallam suggested, by the scepticism of the 1860s. Tennyson, like his age, desperately wanted to believe in the literal reality of the grail, and by implication in the religious certainty it embodied, but he was also open to the increasingly uncomfortable truths of science. Like Darwin with the *Origin*, he had all sorts of excuses for not publishing his key text. They are revealingly contradictory: 'The old writers *believed* in the Grail,' he wrote gloomily to the Duke of Argyll in 1859. However, earlier in the same letter, he had produced the opposite excuse. 'As to Macaulay's suggestion of the Sangreal, I doubt whether such a subject could be handled in these days, without incurring a charge of irreverence.'[32] The uncertainty about his age's response surely suggests both a divided age and a divided poet: how far is Tennyson himself torn between the fear of breaking a religious taboo and the sense that he himself no longer believes in the literal truth of the religious experience? In April 1868 he wrote Ambrosius's 12-line opening speech, questioning Percivale. Then suddenly, in September 1868, after decades of demurring, he wrote the rest of the poem in a fortnight. Hallam quotes from his mother's journal:

> 1868, Sept. 9th. [Alfred] read a bit of his San Graal, which he has just begun.
>
> Sept. 14th. He has almost finished the San Graal. It came like a breath of inspiration.
>
> Sept. 23rd.[Alfred] read the San Graal MS complete in the garden.

Emily charmingly credits the completion of the poem to her own persuasive powers, added to those of (presumably) the Princess Victoria and later, apparently, the Queen herself:

> I doubt whether the San Graal would have been written but for my endeavour, and the Queen's wish, and that of the Crown Princess. Thank God for it. (1660)

There are, of course, more mundane explanations for the delay. One reason might simply be structural: it took Tennyson until 1868 to come upon a solution to the problem of finding a suitable narrative voice. Where Malory, his source, writing in an age of faith, told the story authoritatively, as a third person narrative, Tennyson eventually decides to address the scepticism of the 1860s by telling it through Sir Percivale and then through the voices of the other characters – the holy Nun (Percivale's sister), Sir Galahad, Sir Lancelot and Sir Bors – each of whom has a vision, says Tennyson, adapted to 'their own peculiar natures and circumstances'.[33] The poem is wonderfully polysemic – 'shot-silk with many glancing colours' – and lends itself to being read in many different ways.[34] Hallam Tennyson stresses its religious aspects, praising 'the mystical treatment of every part of the subject'. However, Emily Tennyson's journal, usually a better guide to her husband's thoughts, comments that the poem can be read 'scientifically' as well as spiritually:

> [Alfred] read *The Holy Grail* to the Bradleys, explaining the realism and symbolism and how the natural, if people cared, could always be made to account for the supernatural.[35]

Tennyson seems here to be throwing his lot in with the scientists in implying that there is no final authority for the existence of the grail. Honest Doubt has been satisfied. This shift in meaning entails a structural change to Malory's narrative: unlike Malory, Tennyson keeps King Arthur, the guarantor of authority, away from the central scene where the mystic cup appears: interestingly, it is Arthur, looking back at Camelot from a distance, who gives the most unspiritual 'scientific' explanation of the appearance, seeing it simply as a thunderstorm:

> 'Lo, there! The roofs
> Of our Great Hall are rolled in thundersmoke!
> Pray Heaven they be not smitten by the bolt!'

It is the knights alone, left in Camelot without authority, who succumb to the vision.

> And all at once, as there we sat, we heard
> A cracking and a riving of the roofs,
> And rending, and a blast, and overhead
> Thunder, and in the thunder was a cry.
> And in the blast there smote along the hall
> A beam of light seven times more clear than day:
> And down the long beam stole the Holy Grail
> All over covered with a luminous cloud,
> And none might see who bare it, and it past,
> But every knight beheld his fellow's face
> As in a glory [...][36]

As far as can be seen from this version, Arthur himself never actually *sees* the grail. Far from it being given the stamp of royal approval, its appearance is greeted with dread by King Arthur. At the end of the poem, his apprehension has been amply vindicated, and he indulges in a blank-verse version of 'I told you so!':

> And spake I not too truly, O my knights?
> Was I too dark a prophet when I said
> To those who went upon this Holy Quest,
> That most of them would follow wandering fires,
> Lost in the quagmire?[37]

Hallam adds this: 'He [Tennyson] pointed out the difference between the five visions of the Grail, as seen by the Holy Nun, Sir Galahad, Sir Perceval, Sir

Lancelot, Sir Bors, according to their own peculiar natures and circumstances' (1661). Tennyson deliberately leaves open the possibility of a natural or certainly of a psychological explanation for each sighting. Thus the Nun's vision of the grail is highly sexualized, Galahad's confirms his religious vocation, Sir Percivale's is a plain man's glimpse into things far beyond him. Tennyson is not presenting something he regards as an absolute truth, as Malory did in the fifteenth century, but is exploring adaptation, modification – each individual adapting, finding what they need, from the religious resources available to them. What Tennyson does is to show a sort of spiritual natural selection (a survival of the fittest) where only the spiritually superior, Galahad, sees a clear vision of the grail, and everyone else sees a version adapted to their own limitations. There is no omniscient voice, no divine authority, only the exploratory tones of Sir Percivale, himself another seeker after the grail. Here the spiritual enters the Darwinian world of endless change and adaptation and itself adapts to accommodate the insights of science.

One tiny allusive marker of the presence of the *Origin* behind 'The Holy Grail' is the presence of a blue-eyed cat. Here is Tennyson's casual, throw-away image, uttered by Gawain:

> 'But by mine eyes and by mine ears I swear
> I will be deafer than the blue-eyed cat'[38]

And here is what he may have been remembering, from his own eager reading of *The Origin of Species* in 1859. Darwin observes in chapter 1 of *The Origin*, 'Variation Under Domestication':

> Some instances of correlation are quite whimsical: thus cats which are entirely white and have blue eyes are invariably deaf.[39]

This is not particularly arcane knowledge, of course – Tennyson may well have come across the fact elsewhere – but neither is it exactly the most obvious image to use for deafness. It was in fact Hallam Tennyson who first pointed out a possible link with the *Origin*.[40] It is, though, with all due respect to Hallam, a very minor piece of circumstantial evidence. Here simple chronology suggests a more significant link. What was happening to Tennyson in the weeks before he began work on 'The Holy Grail'? What was it that suddenly released his imagination and enabled him, after all those decades, to begin this crucial poem? In Hallam's memoir is a quotation from Emily's journal for 17 August 1868:

> Aug. 17th Farringford. Mr Darwin called, and seemed to be very kindly, unworldly, and agreeable. [Alfred] said to him, 'Your theory of Evolution does not make against Christianity': and Darwin answered, 'No, certainly not'.[41]

Two weeks later, after decades of demurring, Tennyson wrote 'The Holy Grail' 'in a breath of inspiration'. It is tempting to read this as at least in part deriving from the encounter with Darwin: there was, after all, a way for the grail to be 'true' without the need to reject scientific scepticism – a way to integrate Christian mysticism with Darwinian science. Hence Hallam Tennyson's stress, when reading the poem to the Bradleys, on the constant supply of non-mystical explanations available alongside the mystical ones: 'the natural, if people cared, could always be made to account for the supernatural'.[42]

The final lines of the poem, says Tennyson, are of great significance: 'The end, where the King speaks of his work and of his visions, is intended to be the summing up of all in the highest note by the highest of men'.[43] Arthur, who in Tennyson's version has not seen the grail, returns to Camelot and talks powerfully of his *own* visions.[44] Referring to himself in the third person – another distancing device – the King talks of

' [...] moments when he feels he cannot die,
And knows himself no vision to himself,
Nor the high God a vision, nor that One
Who rose again: *ye have seen what ye have seen.*'
So spake the King: I knew not all he meant.[45]

Percivale here is outside the vision and must use his eyes and ears, as the King reminds him. Faith is no longer presented as blind, but as involving the use of all the faculties, including observation and analysis. 'I knew not all he meant': analysis is crucial, but it may end, like much empirical science, in uncertainty rather than in absolute faith. In 1869, not long after completing 'The Holy Grail', Tennyson helped set up the Metaphysical Society. It was at the inaugural meeting of the Metaphysical Society that T. H. Huxley coined the term 'agnosticism'. I believe that Tennyson's 'Holy Grail' is an early literary manifestation of this slippery but somehow comforting Victorian position, somewhere between belief and scepticism, and that the term itself is linked crucially with one of the implications of Darwin's theory and with what Ricks calls Tennyson's 'art of the penultimate'.[46] Both involve dispensing with certainty and stasis in favour of uncertainty and flux.

The August 1868 meeting was the only one that I have been able to track down between these two men, key figures of their age.[47] What exactly went on? What can we flesh out from Emily's brief account? Was Darwin merely being polite? There is plenty of evidence that his own demurring over publishing *The Origin of Species* was partly to do with his reluctance to destroy people's beliefs. When he did finally publish, he said that he felt as if he were committing a murder. (He was relieved to hear from Charles Kingsley that *he* could read the *Origin*, as he had read *Vestiges*, as still compatible with religious belief.) There is plenty of evidence too of Darwin's own kindliness and modesty, which the perceptive Emily recognized even on this short visit. In what tone of voice did Tennyson say 'Your theory of Evolution does not make against Christianity'? Was it a question? A statement? A defiant assertion? Tennyson uses a similar dialogic structure at the end of 'The Holy Grail', when he returns to that overriding theme of immortality, of an 'Arthur who cannot die' – but this time framed by the puzzled voice of Sir Percivale, through whom he is telling the story. It is a powerfully agnostic ending, melding belief and scepticism, and equally encouraged, I believe, by that visit from Darwin, which contributed to Tennyson's gradual evolution of a sort of Christian agnosticism.

The final reference to Darwin in Tennyson's letters is merely a disparaging reference to biography. Allingham in 1887 asks Tennyson if he has read Darwin's *Life*.[48] Allingham's account is as usual bald but suggestive:

'Have you seen Darwin's Life?'
T. – 'No, I hate biographies.'
Darwin – his dicta on religion – once cared for poetry, etc.[49]

What exactly Tennyson or Allingham meant by Darwin's 'dicta on religion' is unclear. What must undoubtedly have horrified Tennyson was Darwin's move not simply towards agnosticism but away from poetry. According to the *Dictionary of National Biography* (using ample evidence from Darwin's letters), the man who in his youth had considered the church as a profession, and who had responded so powerfully to the English poets, changed as he grew older: 'His literary taste suffered a decay as he grew older – in his youth he found great delight in the poetry of Milton, Shakespeare, Wordsworth, &c. But in later life all such pleasure was dead.'[50]

Tennyson's horror, implied here, at Darwin's fate suggests unexpectedly and counter-intuitively two key similarities between the thought patterns of the two men: the first is that both needed to cultivate and maintain, in their different fields, a sense of mystery and of wonder. Despite the alleged attenuation of his response to literature, it was after all Darwin who borrowed John Herschel's term 'the mystery of mysteries' for his own 'Holy Grail', the source of the origin of new species.[51] George Levine has written movingly of Darwin's continuing capacity for wonder.[52] For Tennyson, the mystery of mysteries and source of wonder was surely his own sense of the continuing presence of Hallam: 'The dead are not dead but alive.'[53]

The second similarity is perhaps the more suggestive: the thinking of both Tennyson and Darwin moves and flows constantly – is inalienably unfixed. Water is the central image in Tennyson's poetry and flux its central trope. Similarly Darwin spoke of his 'fluctuations' of belief, insisting even in his days of declared agnosticism that there were days on which he deserved to be called a theist.[54] John Hedley Brooke goes so far as to suggest that 'Even his atrophied sensibilities were perhaps not as deadened in later life as he pretended.'[55] Darwin shares Tennyson's early fascination with earthquakes and natural cataclysms.[56] He writes: 'A bad earthquake at once destroys our oldest associations: the earth, the very emblem of solidity, has moved beneath our feet like a thin crust over a fluid – one second of time has created in the mind a strange idea of insecurity, which hours of reflection would not have produced.'[57] Fluidity is at the heart of Darwinism and at the heart of Tennyson's poetry too:

> The hills are shadows and they flow
> From form to form and nothing stands.[58]

Meadows's suggestions, cited at the beginning of this chapter, that Tennyson increasingly wondered 'how science and poetry, as he practised it, were to be reconciled', can be addressed very simply. Just as Tennyson towards the end of his life bent towards science, so Darwin in the very nature of his writing, as George Levine has suggested, went on inclining towards poetry. The two men's minds seem to have operated in a strangely similar way, first moving to record and observe, and only then grasping imaginatively. Hence Tennyson's writing out of the long prose versions of the *Idylls* which preceded the poetic enactment of the stories. The shared fluidity of their imaginations undermines any easy opposition between literary and scientific thinking. If 'The Holy Grail', as I have argued, is a Darwinian narrative, then the final paragraph of the *Origin* might equally be said to be Tennysonian.

Notes

1 Algernon Charles Swinburne, 'Dethroning Tennyson: A Contribution to the Tennyson–Darwin Controversy', *Nineteenth Century* 23 (1888): 127–9, cited in Gowan Dawson, *Darwin, Literature and Victorian Respectability* (Cambridge: Cambridge University Press, 2007), 53–4.
2 Cited in *The Poems of Tennyson*, ed. Christopher Ricks (London: Longman, 1969), 1661.
3 Quoted in F. B. Pinion, *A Tennyson Chronology* (Basingstoke: Macmillan, 1990), 87.
4 Cited in *The Life and Letters of Thomas Huxley*, ed. Leonard Huxley, 2 vols (London: Macmillan, 1900), 2:338.

5 This brief discussion owes a great deal to John Holmes's paper 'The Poet of Science: How Scientists Read Their Tennyson', given at the Anglia Ruskin University conference on 'Darwin, Tennyson and Their Readers', 17 October 2009. The material was later included in his monograph *Darwin's Bards* (Edinburgh: Edinburgh University Press, 2009).
6 [Norman Lockyer], unsigned note on the death of Tennyson, *Nature* 46 (1892): 572.
7 Norman Lockyer and Winifred Lockyer, *Tennyson as a Student and Poet of Nature* (London: Macmillan, 1910).
8 A. J. Meadows, 'Astronomy and Geology, Terrible Muses! Tennyson and Nineteenth-Century Science', *Notes and Records of the Royal Society of London* 46 (1992): 116.
9 Ibid., 118.
10 Quoted in Kenneth Clark, *Ruskin Today* (London: Penguin, 1964), 88. Ruskin, of course, went on to reject Darwin, on the same grounds as Samuel Butler.
11 Charles Dickens, *Great Expectations*, chap. 12 (Harmondsworth: Penguin, 1994), 90.
12 Ronald Thomas, 'Wilkie Collins and the Sensation Novel' in *The Columbian History of the British Novel*, ed. John Richeth (New York: Columbia Press, 1994), 506; George Levine, *Darwin the Writer* (Oxford: Oxford University Press, 2011), 137.
13 Quoted in Charles Tennyson, *Alfred Tennyson* (London: Macmillan, 1949), 377.
14 *The Letters of Alfred Lord Tennyson*, ed. Cecil B. Lang and Edgar F. Shannon Jr, 3 vols (Oxford: Clarendon Press, 1981–90), 2:451.
15 Ibid., 450.
16 Eleanor Bustin Mattes, *In Memoriam*, ed. Erik Gray (New York: W. W. Norton, 2004), 139–44.
17 Hallam Tennyson, *Alfred Lord Tennyson: A Memoir by His Son*, 2 vols (London: Macmillan, 1897), 2:23–4.
18 *The Autobiography of Charles Darwin: 1809–1882*, ed. Nora Barlow (London: Collins, 1958), 87.
19 *Poems of Tennyson*, 1022.
20 *Charles Kingsley: His Letters and Memories of His Life*, ed. F. E. Kingsley (London: Kegan Paul, 1883), 146, quoted in John Hedley Brooke, 'Darwin and Victorian Christianity,' in *The Cambridge Companion to Darwin*, 2nd ed., ed. Jonathan Hodge and Gregory Radick (Cambridge: Cambridge University Press, 2009), 213.
21 See 'Despair', *Poems of Tennyson*, 26 and 'Hope', *Poems of Tennyson*, 7–8.
22 George Eliot, *The Mill on the Floss* (London: Penguin, 2003), 543.
23 Graham Hough in John Dixon Hunt, ed., *In Memoriam: A Casebook* (Basingstoke: Macmillan, 1987), 145.
24 *Poems of Tennyson*, 1470 (my italics).
25 'All thoughts, all creeds, all dream, are true', *Poems of Tennyson*, 257–8.
26 'The Passing of Arthur', *Poems of Tennyson*, 1743, line 29.
27 Ibid., 1743 (my italics).
28 'So all day long', *Poems of Tennyson*, 1470; 'There to be healed', *Poems of Tennyson*, 1753.
29 In the *Philological Quarterly* and the *Tennyson Research Bulletin*.
30 The Holy Grail story still has popular resonance in the twenty-first century, as evidenced by the international success of Dan Brown's *The Da Vinci Code* (2003).
31 Quoted in H. Tennyson, *A Memoir*, 1:456.
32 *Poems of Tennyson*, 1661.
33 Ibid.
34 See ibid., 1463.

35 Dated January 1869. Cited in *A Tennyson Chronology*, 120.
36 *Poems of Tennyson*, 1668, lines 219–21.
37 Ibid., 1667–8, lines 182–92.
38 Ibid., 1686, lines 861–2.
39 Charles Darwin, *On the Origin of Species* (London: John Murray, 1859), 12.
40 Ricks observes that Hallam Tennyson himself made the link with the first chapter of the *Origin*, and that he quoted from an early edition: 'Thus cats which are entirely white and have blue eyes are generally deaf; but it has lately been pointed out by Mr. Tait that this is confined to the males' (*Poems of Tennyson*, 1686).
41 Quoted in H. Tennyson, *A Memoir*, 2:57.
42 Cited in F. B. Pinion, *A Tennyson Chronology*, 120.
43 Quoted in *Poems of Tennyson*, 1661.
44 The last four lines of Arthur's speech, said Tennyson, formed the spiritual centre of the whole cycle. H. Tennyson, *A Memoir*, 2:90.
45 My italics.
46 Christopher Ricks, *Tennyson* (London: Macmillan, 1972), 49.
47 Though no further meetings between Tennyson and Darwin have been recorded in detail, there is evidence that some did occur, probably during the Darwins' 1868 stay on the Isle of Wight. Darwin's elder surviving daughter, Henrietta (later Litchfield) recorded of that holiday: 'Tennyson came several times to call on my parents, but he did not greatly charm either my father or my mother.' Emma Darwin, *A Century of Family Letters*, 2 vols (London: John Murray, 1915), 2:190, cited in *Letters of Tennyson*, ed. Lang and Shannon, 2:500. Allingham's *Diary* also records a visit to Farringford, exactly a week before the visit recounted in this chapter, of Darwin's wife, his brother Erasmus and his younger surviving sister, Elizabeth. It seems likely then that Charles Darwin's visit a week later was merely a courtesy call to atone for his absence on the previous occasion.
48 *The Life and Letters of Charles Darwin, Including an Autobiographical Chapter*, ed. Francis Darwin, 3 vols (London: John Murray, 1887).
49 William Allingham, *A Diary* (London: Macmillan, 1907), 367–8, quoted in *Letters of Tennyson*, 3:361.
50 *Dictionary of National Biography*, cited in *Letters of Tennyson*, 3:361.
51 John Hedley Brooke, 'Darwin and Victorian Christianity', in the *Cambridge Companion to Darwin*, ed. Jonathan Hodge and Gregory Radick (Cambridge: Cambridge University Press, 2009), 198.
52 In *Darwin Loves You* (2008) and *Darwin the Writer* (2010) – and in Chapter 8 of the present volume.
53 Alfred Tennyson, 'Vastness', in *Poems of Tennyson*, 1348.
54 See for example, in a letter to John Fordyce, 7 May 1879 (*Darwin Correspondence Project Online*, Letter 12041): 'In my most extreme fluctuations I have never been an atheist in the sense of denying the existence of a God.'
55 Brooke, 204.
56 Notably in 'Timbuctoo', *Poems of Tennyson*, 170–73.
57 Charles Darwin, *Journal of Researches into the Geology and Natural History of the Various Countries Visited by H.M.S. Beagle* (London: Colburn, 1839), 16:369.
58 *In Memoriam* cxxiii, *Poems of Tennyson*, 973.

Chapter 5

'AN UNDUE SIMPLIFICATION': TENNYSON'S EVOLUTIONARY AFTERLIFE*

Michiel Nys

On 30 October 1894, some two years after his father's death, Hallam Tennyson wrote to Thomas Henry Huxley asking him to contribute perhaps a line or two to the official biography he was preparing. Not long before his final illness, Hallam had taken his aged father to the Natural History Museum in London, where they had seen Boehm's statue of Charles Darwin, which Huxley had officially inaugurated there.[1] He now suggested that Huxley might provide a critical estimate of the late Poet Laureate's outlook on religion, on science and on the soul – a rather broad assortment of topics which, Hallam must have hoped, Huxley was uniquely placed to weave together and fashion into a coherent picture of Tennyson's attitude to the matters of life and death. After all, these were subjects upon which both men had published and which they had also discussed together at the Metaphysical Society in the 1870s.[2]

Thomas Huxley himself had collapsed once already, during the previous winter. His health had picked up in the spring, and he had polished off a somewhat similar piece for *The Life of Richard Owen*, undertaken at the request of Sir Richard's grandson. In addition, Huxley had finished the 'Prolegomena' to 'Evolution and Ethics', with which he prefaced the text of his 1893 Romanes lecture in the ninth volume of his *Collected Essays*. Taken together, this essay and its accompanying 'Prolegomena' contained Huxley's own definitive answer to the questions implicit in Hallam Tennyson's request. His health was fine when he promised Hallam to look into Tennyson's attitudes to mortality. But just a couple of months later, Huxley's fatal illness set in. His personal estimate of

---

* I would like to thank Dr Ortwin de Graef for reading and providing insightful commentary on a draft version of this chapter.

the poet was never written. Unlike John Tyndall's, we do not find it included in Hallam's *Memoir*.

T. H. Huxley's appreciation of Tennyson's poetry is well known. 'Westminster Abbey', the poem which he wrote on Tennyson's funeral, is instructive for the conventional piety with which Huxley described Tennyson taking his place among

> The men of state, the men of song;
> The men that would not suffer wrong;
> The thought-worn chieftains of the mind,
> Head servants of the human kind.[3]

'Westminster Abbey' rather plainly fed into prevailing conceptions of national leadership and ideal continuity down the generations.

> Bring me my dead!
> To me that have grown
> Stone laid upon stone,
> As the stormy brood
> Of English blood
> Has waxed and spread
> And filled the world,
> With sails unfurled;
> With men that may not lie;
> With thoughts that cannot die.[4]

The poem describes a private emotional sacrifice made by the individual – stolen, as it were, from nature, and the immediate circle of the family – for the benefit of the entire nation, and perhaps even humanity at large.

> And oh! sad wedded mourner, seeking still
> For vanished hand-clasp; drinking in thy fill
> Of holy grief: forgive, that pious theft
> Robs thee of all, save memories, left:
> Not thine to kneel beside the grassy mound
> While dies the western glow; and all around
> Is silence; and the shadows closer creep
> And whisper softly: All must fall asleep.[5]

A comparison with Charles Darwin's burial at the Abbey ten years earlier readily suggests itself, and the fact that Matthew Arnold published an elegy with the same title on Arthur Penrhyn Stanley's death also suggests something of Huxley's readiness to perpetuate, poetically, an accepted tradition of national mourning. Writing to his wife, Huxley proudly compared his effort – stylistically 'hammered' but emotionally 'human' – to others assembled in the November 1892 issue of the *Nineteenth Century*. True, they were 'castings of much prettier pattern', but, he added, 'I do not think there is a line of mine one of my old working-class audience would have boggled over'.[6]

George John Romanes, though he could hardly be called working-class, was so moved by Huxley's effort that he sent him a copy of his own privately published poems.[7] Emotional investment in Tennyson's poetry seems to have been extensive in Huxley's immediate circle. On one memorable occasion, Charles Darwin's own lack of reverence for the Poet Laureate led Huxley's wife to compare Darwin – in one go – to both Richard Owen and the Bishop of Wilberforce, those arch-villains of Darwinian mythology.[8] Anecdotes like these, as they come down to us largely from the venerable tradition of lives and letters, bring into relief the often intricate interweaving of private emotional investment with the larger spheres of Victorian public discourse and later cultural memory.[9]

What I want to focus on in this essay is Tennyson's legacy in the field of biology, summed up as it usually is in that one famous phrase from *In Memoriam*, 'Nature, red in tooth and claw'.[10] If, as Hallam's request to Huxley implied, Tennyson's metaphysical outlook or position is at stake, such a late poem as 'The Making of Man' can be taken as a rather straightforward measure of the extent to which the poet infused the idea of organic evolution with dualistic and teleological-eschatological notions. These were largely derived, of course, from Christianity, and were within the purview not of the man of science but of the prophet:

> Where is one that, born of woman, can altogether escape
> From the lower world within him, moods of tiger, or of ape?
> Man as yet is being made, and ere the crowning Age of ages,
> Shall not aeon after aeon pass and touch him into shape?
>
> All about him shadow still, but, while the races flower and fade,
> Prophet-eyes may catch a glory slowly gaining on the shade,
> Till the peoples all are one, and all their voices blend in choric
> Hallelujah to the Maker 'It is finished. Man is made.'[11]

Tennyson drew for his vision on evolution and its materiality – the universal condition of being 'born of woman' – but located a supreme realization of it in the spirit of 'the Maker'. In this respect, nature, and the flowering and fading races of the earth (including presumably 'the stormy brood / Of English blood'), fell decidedly short.[12] Needless to say, Tennyson's take on the matter was essentially different from T. H. Huxley's. Yet Huxley alluded to Tennyson's poetry on a handful of occasions, and their names have regularly been coupled.[13]

In late July 1918 the young Aldous Huxley, not quite 24 at the time, sent a poem to his elder brother, the biologist Julian Huxley, who was then at the British Mission in Padua, serving as a lieutenant in the Intelligence Corps. Aldous described his production as 'the most lovely little song [...] – quite Tennysonian, both in respect to its perfection of form and in its recognition, so highly acclaimed by Grand Pater in the late Poet Laureate, of the truths of Science'. In addition, it was rather amusing.

> A million million spermatozoa
> All of them alive:
> Out of their cataclysm but one poor Noah
> Dare hope to survive.
>
> And among that billion minus one
> Might have chanced to be
> Shakespeare, another Newton, a new Donne;
> But the One was Me.
>
> Shame on you to oust your betters thus,
> Take ark leaving the rest outside!
> Better for all of us, forward Homunculus,
> If you'd quietly died.[14]

Rather like Tennyson in 'The Making of Man', Aldous Huxley highlighted the curious, almost laughable deficiency of material nature, though unlike Tennyson, he drew no comfort from any suggestion of the potential immortality of the soul. This particular version of the struggle for existence – a war poem by a man rejected from military service, after all – seemed to puncture heroic Victorian efforts at relating history at the levels of the individual, the family and the nation with history on a global, and biological, scale.

Yet that seemingly Victorian enterprise is largely what the poem's recipient was to make his career. In 1944, Julian Huxley, who had by then become a familiar voice on BBC radio, and one of the most illustrious biologists of his day, published a volume of essays titled *On Living in a Revolution*. Taken together, these essays argued that the evils of war and totalitarianism could be subsumed under a more far-reaching revolution. After the War, history was to bring about a peaceful, liberal and democratic world order. Included in the volume was a transcript of a BBC radio dramatisation broadcast in October 1942 as 'Thomas Henry Huxley and Julian Huxley: An Imaginary Interview'. Heavily didactic in tone, the programme had Julian explain to a fictional T. H. that the twentieth century had embraced moral relativism in a way that the nineteenth never had. The ensuing lack of certainty, however, had made it hard to deal squarely with the ideological threat posed by fascism. In the absence of traditional Christianity, any positive, meaningful, encouraging alternative to an Italian- or German-style national religion seemed to be wanting. Consequently, the problem of cosmic justice, in which T. H. Huxley had taken such a special interest at the end of his life, had again become acute:

> JULIAN: But there's a question which I have longed to ask you ever since, as a young man, I read your famous Romanes lecture, *Evolution and Ethics*. There you stated (I remember the passage vividly) that the ethical progress of society depends not on imitating the cosmic process but in combating it, and by the cosmic process you of course meant mainly the ruthless struggle for existence. As an evolutionist, I never understood how man, himself a part of nature, could fulfil his destiny by fighting against that same process which gave him birth.
>
> THOMAS HENRY: Is it not self-evident? Any theory of ethics cannot but repudiate the gladiatorial theory of life; the practice of virtue must be opposed to the type of conduct which is successful in the cosmic struggle for existence.[15]

True enough, T. H. Huxley's Romanes lecture had elicited a certain amount of commentary when the Second World War broke out. In mid-December 1939, the ruralist writer H. J. Massingham, for instance, found himself caught up in a heated debate over its significance in the letters pages of the *Times*. Massingham deplored, from a quasi-mystic, utopian nostalgic perspective, T. H. Huxley's coupling of nature and struggle, civilization and peace:

> That Huxley compared 'the sighs and groans of pain' in Nature to what Dante saw 'at the Gate of Hell' reveals how far he did go. [...] If this unbalanced charge were well founded, Nature would hardly have been the inspiration of our poetry (the greatest in the world) for so many centuries, nor should we have seen in the pursuit of husbandry – wherein Nature and man interact more intimately than in any other of man's practical activities – that profound satisfaction of life upon which the benediction of Christ rested so frequently in the Parables. All art, as Shakespeare saw, derives ultimately from Nature, as it could not have done if rapacity had been the sole motive force of natural law. The ugliness of cruelty and greed is itself the answer.[16]

Though he would hardly have agreed with Massingham, Julian Huxley, too, felt the urge to absolve nature – or, properly speaking, evolution – from any responsibility for the horrors of human warfare:

> THOMAS HENRY (reminiscently): The struggle for existence – my friend Tennyson summed it up: 'Nature red in tooth and claw.'
>
> JULIAN: That appears to have been an undue simplification.[17]

And Julian went on to point out, first, the role of intelligence and cooperation; second, the difference between individual selection and group selection (which he considered to be a refutation of biologically inspired theories of economic laissez-faire); and, finally, as he saw it, 'the demonstration that there *is* such a thing as progress in biological evolution', which takes the form of 'increased harmony of construction, increased capacity for knowledge and for feeling, and increased control over nature, increased independence of outer change'.[18]

It was a subject which Julian Huxley pursued throughout his career, often with reference to his grandfather's Romanes lecture. In an earlier book, *Essays in Popular Science*, for instance, he had included a critical discussion of the lecture.[19] The zoologist H. M. Parshley, in the book section of the *New York Herald Tribune*, agreed wholeheartedly with Julian's criticisms – even if their author 'lacks something of the vigor and beauty of style that characterize his ancestor's work' – and went so far as to improve upon it:[20]

> To such an absurdity can innate Puritanism, a life of struggle, and a fundamentally 'reconciling' disposition lead a spirit of supreme integrity! Combat the cosmic process, indeed! The one great ethical principle that we now see to be firmly based on evolutionary, biological science is that success and survival as well as physical and spiritual freedom absolutely depend upon adaptation to the cosmic process, knowledge of and obedience to its laws, and alliance with it. Our author, of course, disagrees with Huxley's principle; and he shows how it was possible to be enunciated, even by one who believed in the evolutionary origin of both good and bad impulses.

Julian Huxley even incorporated the problem – and its original Tennysonian wording – in his scientific magnum opus, *Evolution: The Modern Synthesis*:

> The poet spoke of letting ape and tiger die. To this pair, the cynic later added the donkey, as more pervasive and in the long run more dangerous. The evolutionary biologist is tempted to ask whether the aim should not be to let the mammal die within us, so as the more effectually to permit the man to live.[21]

He developed his thesis in his own Romanes lecture for 1943, delivered exactly fifty years after his grandfather's.[22]

Julian Huxley's almost Spencerian version of evolution did not survive for very long. In the opening chapter to *The Selfish Gene*, Richard Dawkins programmatically set up his argument in opposition to the group-selectionist models found in such older books of popular science as Konrad Lorenz's *On Aggression*, Robert Ardrey's *The Social Contract*, and Irenäus Eibl-Eibesfeldt's *Love and Hate* – roughly inspired by the scientific literature Julian Huxley had promoted (by W. C. Allee, for instance, and Alfred Emerson), and drawn on for support. Dawkins described his project very much as a return to the original Darwinian understanding of 'Nature, red in tooth and claw':

> It is ironic that Ashley Montagu [who was a student of Julian Huxley in the mid-1920s] should criticize Lorenz as a 'direct descendant of the 'nature red in tooth and claw' thinkers of the nineteenth century...'. As I understand Lorenz's view of evolution, he would be very much at one with Montagu in rejecting the implications of Tennyson's famous phrase. Unlike both of them, I think 'nature red in tooth and claw' sums up our modern understanding of natural selection admirably.[23]

Interestingly, Dawkins associated the tooth-and-claw perspective, just as Julian Huxley had done before him, with the outlook defended by T. H. Huxley, specifically in his Romanes lecture.[24]

Let us now look at T. H. Huxley's own position, both scientific and moral, and examine when and how he explicitly drew on Tennyson's verse for support. In fact, Huxley hailed Tennyson as *the* poet of modern science even before Huxley was an evolutionist, before he knew of Darwin's views on natural selection, in the 1850s. What is more, Huxley first invoked the scientific authority of Tennyson not to reinforce but to repudiate the tooth-and-claw outlook as he knew it in biology.

On 15 February 1856, T. H. Huxley delivered a Friday evening lecture at the Royal Institution. In earlier work, he had already made clear his opposition to theories of what he now called 'the possibly fortuitous development of animal life'.[25] But in this particular lecture, the developmental hypothesis was not really at issue. Indeed, he now opposed the followers of the French comparative anatomist and palaeontologist Georges Cuvier, who had himself always challenged the varieties of evolutionism propounded by his compatriots Jean-Baptiste Lamarck, and Geoffroy Saint-Hilaire. In particular, Huxley targeted the British popularizers of 'the prince of modern naturalists', the natural theologians and the advocates of the Frenchman's so-called laws of physiological correlation.[26] The idea found its supreme illustration in the anatomy and physiology of the vertebrate carnivores – where, it was sometimes claimed, finding a single fossil bone sufficed for the knowledgeable palaeontologist to deduce any and every detail of the anatomy of the animal:

> In a word, the form of the tooth involves that of the condyle: that of the shoulder blade; that of the claws: just as the equation of a curve involves all its properties. And just as by taking each property separately and making it the base of a separate equation, we should obtain both the ordinary equation, and all other properties whatsoever which it possesses; so, in the same way, the claw, the scapula, the condyle, the femur, and all the other bones taken separately will give the tooth, or one another; and by commencing with any one, he who had a rational conception of the laws of the organic economy, could reconstruct the whole animal.[27]

Cuvier might claim he came by his fossil reconstructions by necessary deduction, moving from an incomplete fossil to the autonomous, physiologically functional organism and its particular habits of diet and life. To Huxley, however, this was a misrepresentation of the empirical and inductive processes of reasoning involved. Cuvier, Huxley asserted rather categorically, 'did not himself understand the methods by which he arrived at his great results'.[28] And this led to a radically distorted view of nature – a perpetually recurrent hypothesis (utility from the point of view of the organism, adaptation of the organism to its particular way of life) which, given the necessary absence of any evidence to the contrary, was always self-fulfilling. Huxley asked, rhetorically:

> Is this utilitarian adaptation to a benevolent purpose the chief, or even the leading feature of that great shadow, or, we should more rightly say, of that vast archetype of the human mind, which everywhere looms upon us through nature? The reply of natural history is clearly in the negative. She tells us that utilitarian adaptation to purpose is not the greatest principle worked out in nature, and that its value, even as an instrument of research, has been enormously overrated.[29]

Huxley rejected any purposive drive in nature as the result of anthropomorphic projection: 'In the words of the only poet of our day who has fused true science into song, the philosopher, looking into Nature, "Sees his shadow glory-crowned, / He sees himself in all he sees."'[30] This is the moral he derived from *In Memoriam*. Not all mammals fit the carnivore-herbivore dichotomy, and when dealing with lower organisms, the idea of physiological necessity seemed wholly inapplicable. How did the principle of physiological correlation apply to fossil plants? What use was served by the beauty and unity in diversity of plant and animal form?

> Who has ever dreamed of finding an utilitarian purpose in the forms and colours of flowers, in the sculpture of pollen-grains, in the varied figures of the fronds of ferns? What 'purpose' is served by the strange numerical relations of the parts of plants, the threes and fives of monocotyledons and dicotyledons?
>
> Thus in travelling from one end to the other of the scale of life, we are taught one lesson, that living nature is not a mechanism but a poem; not a mere rough engine-house for the due keeping of pleasure and pain machines, but a palace whose foundations, indeed, are laid on the strictest and safest mechanical principles, but whose superstructure is a manifestation of the highest and noblest art.[31]

The point of Huxley's lecture was not simply to fix the methodology but to define the rationale for studying natural history as well. This, too, was open-ended:

> Science, as power, indeed, showers daily blessings upon our practical life; and science, as knowledge, opens up continually new sources of intellectual delight. But neither knowing nor enjoying are the highest ends of life. Strength – capacity of action and of endurance – is the highest thing to be desired; and this is to be obtained only by careful discipline of all the faculties, by that training which the pursuit of science is, above all things, most competent to give.[32]

Thus, the supreme worth of science was, in the first place, moral. And the touchstone of morality, including the morality imbued by science, was not utility, but attention – attention to every anatomical detail, to the singularity of each and every organism:

> Let those who doubt the efficacy of science as a moral discipline make the experiment of trying to come to a comprehension of the meanest worm or weed, of its structure, its habits, its relation to the great scheme of nature. [...] There is not one person in fifty whose habits of mind are sufficiently accurate to enable him to give a truthful description of the exterior of a rose.[33]

Hugh Falconer, a botanist and geologist who had recently left a government-sponsored position in India to spend the rest of his career doing fossil work in England and continental Europe, attacked Huxley's lecture in the June number of the *Annals and Magazine of Natural History*.[34] Falconer was bewildered by Huxley's preference of empirical over rational explanation, by his using the moral term, utilitarian, to describe the position which he intended to call into question, as well as by the replacement which Huxley had proposed: 'Let him be the great expounder of its aesthetics, if he likes – every one will cheer him on. But he must beware of attempting to put back the hand of the rational dial, for every arm will be against him.'³⁵ Falconer thought Huxley perverse in switching the topic to plants and lower organisms. But he deliberately missed the point of Huxley's empirical purism. In order for life to be understood scientifically, it had to be represented aesthetically; in order for life to be represented aesthetically, it had to be lifted out of life as, in palaeontology, or in art, it usually was. Huxley's reply was printed in the July number of the *Annals and Magazine of Natural History*:

> Let us imagine that all existing animals had perished, but that their dead forms were gathered together and submitted to the investigation of some intelligent being from whom the knowledge that they had ever lived was concealed. [...] He would not term Lions and Tigers and Wolves 'Carnivora,' for he would not even know that they eat anything, but he would assuredly form a group with pretty nearly the same limits as the Carnivora, simply because all these animals resemble one another, and differ from the rest in certain peculiarities of dentition, &c.³⁶

Huxley was egged on in the dispute by Joseph Dalton Hooker, who relayed the details to Charles Darwin.³⁷ Darwin's reaction was entirely characteristic: 'I think Huxley's argument best. — But to deny all reasoning from adaptation & so called final causes, seems to me preposterous. But I am most heartily sorry at the whole dispute: it will prevent two very good men from being friends.'³⁸

In the first instance, then, Huxley did not turn to Tennyson to find poetic justification for any pre-existing scientific conception of 'Nature, red in tooth and claw'; rather, he used Tennyson's verse as a text in order to enlarge on the primarily descriptive task of the scientific expert, challenging what he himself at least saw as the established interpretive orthodoxy. When he delivered his Romanes lecture at Oxford, in 1893, Huxley again invoked Tennyson's poetry in support of his own conceptions of nature and morality.

In the spring of 1893, T. H. Huxley was putting the finishing touches to a lecture he wrote and rewrote again and again, trying as best he could to condense it. He was recovering slowly from influenza contracted in March. In mid-April, Benjamin Jowett congratulated him on his choice of topic: 'No one has yet expressed adequately the antithesis of the moral & the physical.'³⁹ Just a week later, Huxley wrote to Romanes with misgiving: 'If the whole thing is too much for the Dons' nerves – I am no judge of their delicacy – I am quite ready to give up the lecture.'⁴⁰ But lecturing on ethics at Oxford, he thought, was 'decidedly the most piquant occurrence in my career', and he expressed himself as carefully and discreetly as he could.⁴¹ On May 18, at the Sheldonian Theatre, he read the essay out.⁴² He knew that suffering and consciousness had evolved together; that no aboriginal sin or merit was involved. But humanity had risen up. In Tennyson's terms:

> Arise and fly
> The reeling Faun, the sensual feast;
> Move upward, working out the beast,
> And let the ape and tiger die.[43]

Pity, sympathy and care had taken the place of ruthless competition though, Huxley emphasized, the triumph of ethics was hard-fought. A phrase from *In Memoriam* brought out, in particularly memorable form, the ambivalence at the heart of his vision:

> After the manner of successful persons, civilized man would gladly kick down the ladder by which he has climbed. He would be only too pleased to see 'the ape and tiger die.' But they decline to suit his convenience; and the unwelcome intrusion of these boon companions of his hot youth into the ranged existence of civil life adds pains and griefs, innumerable and immeasurably great, to those which the cosmic process brings on the mere animal.[44]

Huxley's refusal to derive any moral lessons from the book of nature, beside the vital need to understand it, in its every detail, and to work with it, confused some of his contemporaries. Herbert Spencer was irritated, St George Mivart elated.[45] But it was in line with the scientific and moral outlook Huxley had expressed since the 1850s. It is here that we expect the oft-quoted section LVI to come in.

> 'Thou makest thine appeal to me:
> I bring to life, I bring to death:
> The spirit does but mean the breath:
> I know no more.' And he, shall he,
>
> Man, her last work, who seem'd so fair,
> Such splendid purpose in his eyes,
> Who roll'd the psalm to wintry skies,
> Who built him fanes of fruitless prayer,
>
> Who trusted God was love indeed
> And love Creation's final law –
> Tho' Nature, red in tooth and claw
> With ravine, shriek'd against his creed –

> Who loved, who suffer'd countless ills,
> Who battled for the True, the Just,
> Be blown about the desert dust,
> Or seal'd within the iron hills?[46]

Instead, Huxley ended his lecture, borrowing, once more, from Tennyson, as if to answer the poet's query in his own terms:

> To strive, to seek, to find, and not to yield,

cherishing the good that falls in our way, and bearing the evil, in and around us, with stout hearts set on diminishing it. So far, we all may strive in one faith towards one hope:

> It may be that the gulfs will wash us down,
> It may be we shall touch the Happy Isles,
>
> [...] but something ere the end,
> Some work of noble note may yet be done.[47]

In the original version, Tennyson's poem then proceeds: 'Not unbecoming men who strove with Gods.'[48] Hallam Tennyson, then, may not have been able to retrieve any notes on the piece which Huxley had promised him before he died, but, in a sense, Huxley had dealt with the topic through quotation and allusion.[49]

Richard Dawkins called on Huxley's Romanes lecture to buttress his peculiar variety of human exceptionalism:

> We have the power to defy the selfish genes of our birth and, if necessary, the selfish memes of our indoctrination. We can even discuss ways of deliberately cultivating and nurturing pure, disinterested altruism – something that has no place in nature, something that has never existed before in the whole history of the world. We are built as gene machines and cultured as meme machines, but we have the power to turn against our creators. We, alone on earth, can rebel against the tyranny of the selfish replicators.[50]

George Williams, too, declared Huxley had been right all along.[51] It is to a version of Huxley's substantially pre-Darwinian moral outlook, then, that heroic neo-Darwinists of the later twentieth century have turned.

Huxley's scientific focus in 'Evolution and Ethics' still led him away from utilitarian adaptation – a process he referred to, if at all, to repudiate theories of inevitable progress. Huxley's emphasis was not on the division of labour in the natural economy. Instead, he drew attention to developmental morphology – illustrated by the life cycle of a bean – and the science of heredity, which had helped to understand the unity in diversity he still admired:

> Cosmic evolution may teach us how the good and the evil tendencies of man may have come about; but, in itself, it is incompetent to furnish any better reason why what we call good is better than what we call evil than we had before. Some day, I doubt not, we shall arrive at an understanding of the aesthetic faculty; but all the understanding in the world will neither increase nor diminish the force of the intuition that this is beautiful and that is ugly.[52]

Beyond this old dichotomy (see Massingham's contention that 'The ugliness of cruelty and greed is itself the answer'), it seemed, there was little that those who 'talk'd with rocks and trees' could say or explain to those who did not.

> Her faith is fixt and cannot move,
> She darkly feels him great and wise,
> She dwells on him with faithful eyes,
> 'I cannot understand: I love.'[53]

As one Tennysonian evolutionist had it:

> The mind of the student of Nature is apt to form the habit of looking upon human life as a spectacle [...] as evanescent as the picture the moon looked down upon during the ages that produced the coal-formations. Original temperament, however, has no doubt a good deal to do with this mood[...][54]

As he wrote this, Theodore Watts was, in fact, contrasting Charles Darwin's notorious, iconic disaffection with poetry in later life with Huxley's poetic affinities as displayed so bountifully in the Romanes lecture. But his remark reflects, perhaps more suggestively, the diverse ways in which, as Gillian Beer has emphasized so persuasively, various evolutionists since Darwin have chosen to frame their biological outlook, embodying it in the aboriginal language of poetry, value and emotion.[55]

Notes

1 Michael Thorn, *Tennyson* (London: Abacus, 1993), 518.
2 Warren R. Dawson, *The Huxley Papers: A Descriptive Catalogue of the Correspondence, Manuscripts and Miscellaneous Papers of The Rt. Hon. Thomas Henry Huxley, P.C., D.C.L.,*

F.R.S., preserved in the Imperial College of Science and Technology, London (London: Macmillan, 1946), 152.
3 Thomas Henry Huxley, 'Westminster Abbey', in Henrietta A. Huxley, *Poems of Henrietta A. Huxley with Three of Thomas Henry Huxley* (London: Duckworth, 1913), 4.
4 Ibid., 3.
5 Ibid., 4. Tennyson's most recent biographer mentions that Thomas Hardy 'characteristically thought a country churchyard would have been more appropriate' (Thorn, *Tennyson*, 525).
6 Thomas Henry Huxley, *Life and Letters of Thomas Henry Huxley*, ed. Leonard Huxley, 2 vols (New York: Appleton, 1900), 2:360. The founding of *Nineteenth Century* by James Knowles in March 1877 was, incidentally, a direct corollary of the discussions of the Metaphysical Society. Both Huxley and Tennyson took part. See Alan Willard Brown, *The Metaphysical Society: Victorian Minds in Crisis, 1869–1880* (New York: Octagon Books, 1973), 185.
7 T. H. Huxley, *Life and Letters*, 2:360.
8 Henrietta Huxley to Charles Darwin, 1 January 1865, Darwin Correspondence Project Database, American Council of Learned Societies and University of Cambridge: www.darwinproject.ac.uk/entry-4733 (accessed 30 May 2011).
9 On the Victorian tradition of scientific biography, see Janet Browne, 'The Charles Darwin–Joseph Hooker Correspondence: An Analysis of Manuscript Resources and Their Use in Biography', *Journal of the Society for the Bibliography of Natural History* 8, no. 4 (1978): 351–66.
10 Gillian Beer, 'Lineal Descendants: The *Origin*'s Literary Progeny', in: *The Cambridge Companion to the 'Origin of Species'*, ed. Michael Ruse and Robert J. Richards (Cambridge: Cambridge University Press, 2009), 275–94; Michael Ruse, *The Darwinian Revolution: Science Red in Tooth and Claw* (Chicago and London: University of Chicago Press, 1999), 150–52.
11 Alfred Tennyson, 'The Making of Man', in *The Poems of Tennyson*, ed. Christopher Ricks, 2nd ed., 3 vols (Harlow: Longman, 1987), 3:248–9.
12 See also: Joseph Warren Beach, *The Concept of Nature in Nineteenth-Century English Poetry* (Russell and Russell: New York, [1936] 1966), 406–34; Tess Cosslett, *The 'Scientific Movement' and Victorian Literature* (Brighton: Harvester Press, 1982), 74–100.
13 For an extended discussion of Tennyson's conception of evolution as it emerges from his last poems, see John Holmes, *Darwin's Bards: British and American Poetry in the Age of Evolution* (Edinburgh: Edinburgh University Press, 2009), 62–74. On his reputation as the most scientific of Victorian poets, see especially 62–3.
14 Grover Smith, ed., *Letters of Aldous Huxley* (London: Chatto & Windus, 1969), 158.
15 Julian Huxley, 'Thomas Henry Huxley and Julian Huxley: An Imaginary Interview', in *On Living in a Revolution* (London: Chatto & Windus, 1944), 83–9 (86).
16 Harold J. Massingham, 'Conflict with Nature', *Times* (2 December 1939), 4.
17 J. Huxley, 'Thomas Henry Huxley and Julian Huxley', 87. Julian Huxley wrote a number of popular pieces emphasizing that, of all animals, ants and humans were the only ones actually to wage organized wars, e.g. 'War as a Biological Phenomenon', in *On Living in a Revolution*, 60–68.
18 J. Huxley, 'Thomas Henry Huxley and Julian Huxley', 87–8. On Julian Huxley's notion of progress, see Marc Swetlitz, 'Julian Huxley and the End of Evolution', *Journal of the History of Biology* 28, no. 2 (1995): 181–217 and Robert M. Gascoigne, 'Julian Huxley and Biological Progress', *Journal of the History of Biology* 24, no. 3 (1991): 433–55.
19 Julian Huxley, 'Thomas Henry Huxley and Religion', in *Essays in Popular Science* (London: Penguin, 1937), 119–38.

20 H. M. Parshley, 'Man, Theology and the Tadpole', *New York Herald Tribune Books* (20 February 1927), 7.
21 Julian Huxley, *Evolution, the Modern Synthesis*, the definitive edition, with a new foreword by Massimo Pigliucci and Gerd B. Müller (Cambridge, MA and London: MIT Press, 2010), 575.
22 Julian Huxley, 'Evolutionary Ethics', in Thomas Henry Huxley and Julian Huxley, *Evolution and Ethics, 1893–1943* (London: Pilot Press, 1947), 103–52.
23 Richard Dawkins, *The Selfish Gene*, 30th anniversary ed. (Oxford and New York: Oxford University Press, 2006): 2. For a summary discussion of Dawkins's position, see John Dupré, *Darwin's Legacy: What Evolution Means Today* (Oxford: Oxford University Press, 2005), 21–2.
24 Richard Dawkins, 'A Devil's Chaplain', in *A Devil's Chaplain: Selected Essays by Richard Dawkins*, ed. Latha Menon (London: Phoenix, 2004), 12.
25 Thomas Henry Huxley, 'On Natural History as Knowledge, Discipline and Power', in *The Scientific Memoirs of Thomas Henry Huxley*, 5 vols, ed. Michael Foster and E. Ray Lankester (London: Macmillan, 1898–1903), 1:305–14 (306).
26 Ibid., 307.
27 Ibid., 308.
28 Ibid.
29 Ibid., 307.
30 Ibid.
31 Ibid., 311.
32 Ibid., 305.
33 Ibid., 313–14.
34 Adrian Desmond, *Huxley: The Devil's Disciple* (London: Michael Joseph, 1994), 227.
35 Hugh Falconer, 'On Prof. Huxley's Attempted Refutation of Cuvier's Laws of Correlation, in the Reconstruction of Extinct Vertebrate Forms', *The Annals and Magazine of Natural History* 17 (1856): 476–93 (493).
36 Thomas Henry Huxley, 'On the Method of Palaeontology', in *Scientific Memoirs*, 1:432–44 (435).
37 Leonard Huxley, *Life and Letters of Joseph Dalton Hooker*, 2 vols (London: John Murray, 1918), 1:426–7.
38 Charles Darwin to Joseph Dalton Hooker, 17–18 June 1856, Darwin Correspondence Project Database, American Council of Learned Societies and University of Cambridge: www.darwinproject.ac.uk/entry-1904 (accessed 30 May 2011).
39 Quoted in Adrian Desmond, *Huxley: Evolution's High Priest* (London: Michael Joseph, 1997), 214.
40 Quoted in T. H. Huxley, *Life and Letters*, 2:376.
41 Ibid., 2:377.
42 Ibid., 2:378–9.
43 Alfred Tennyson, *In Memoriam A.H.H.*, in *Poems of Tennyson*, 2:439.
44 Thomas Henry Huxley, *Evolution and Ethics*, edited and with a new introduction by Michael Ruse (Princeton and Oxford: Princeton University Press, 2009), 52.
45 Desmond, *Huxley: Evolution's High Priest*, 216–17.
46 Tennyson, *In Memoriam*, in *Poems of Tennyson*, 2:373.
47 T. H. Huxley, *Evolution and Ethics*, 86.
48 Tennyson, 'Ulysses', in *Poems of Tennyson*, 2:619.
49 Dawson, *Huxley Papers*, 152.

50 Dawkins, *Selfish Gene*, 201.
51 George C. Williams, 'Mother Nature Is a Wicked Old Witch', in *Evolutionary Ethics*, ed. Matthew H. Nitecki and Doris V. Nitecki (New York: State University of New York Press, 1993), 217–31 (217, 227–30).
52 T. H. Huxley, *Evolution and Ethics*, 80.
53 Tennyson, *In Memoriam*, in *Poems of Tennyson*, 2:417.
54 Theodore Watts[-Dunton], 'Aspects of Tennyson. VI. As the Poet of Evolution', *The Nineteenth Century* 34, no. 200 (October 1893): 657–72 (660).
55 See Gillian Beer, *Darwin's Plots: Evolutionary Narrative in Darwin, George Eliot and Nineteenth-Century Fiction*, 2nd ed. (Cambridge: Cambridge University Press, 2000), xviii–xxx. George Levine, *Darwin Loves You: Natural Selection and the Re-enchantment of the World* (Princeton: Princeton University Press, 2006) is, of course a deliberate programmatic incursion into this tradition.

Chapter 6

'LIKE A MEGATHERIUM SMOKING A CIGAR': DARWIN'S *BEAGLE* FOSSILS IN NINETEENTH-CENTURY POPULAR CULTURE

Gowan Dawson

'There is nothing like geology,' Charles Darwin wrote to his sister Catherine from the Falkland Islands in April 1834. Even for someone who had spent so much of his youth enjoying the bloody thrills of traditional field sports, 'the pleasure of the first days partridge shooting or first days hunting cannot be compared to finding a fine group of fossil bones, which tell their story of former times with an almost living tongue.'[1] Beginning at Punta Alta in September 1832 and continuing at a variety of locations into early 1834, Darwin spent much of the South American leg of the global voyage of HMS *Beagle* eagerly collecting the fossilized remains of extinct prehistoric creatures. As he wrote in his letter to Catherine, uncovering these gigantic osseous remains was often the source of intense pleasure, and he told his other sister, Caroline, in October 1832, 'I have been wonderfully lucky, with fossil bones. — some of the animals must have been of great dimensions: I am almost sure that many of them are quite new; this is always pleasant, but with the antediluvian animals it is doubly so.'[2] The 'cargoes of apparent rubbish' which Darwin continually brought aboard sometimes occasioned 'smiles' of a different sort from his more sceptical shipmates, as the captain, Robert FitzRoy, later recalled, but Darwin was largely correct in anticipating the novelty and importance of many of his finds.[3] Even as the colossal bones were being extricated from the muddy river banks in which they were encased, he attributed many of them, at least provisionally, to the genus *Megatherium*. The megatherium had first been described by the French anatomist Georges Cuvier in 1796 from engravings of an incomplete skeleton in the Royal Museum at Madrid and was already well known in Britain, amongst both specialists and the wider public, despite the scarcity of actual specimens before the early 1830s.

Darwin's discovery of megatherium remains in faraway South America, which would provide some of the specimens so badly needed by British savants, also helped to transmit his burgeoning sense of geological pleasure back across the Atlantic. His sister Susan wrote in March 1833, 'I congratulate you on your luck in finding those curious remains of the Monster M—I think Geology far the most interesting subject one can imagine,' to which her brother replied, 'I am quite delighted to find, the hide of the Megatherium has given you all some interest in my employments.'[4] As well as the putative megatherium hide described in Darwin's letters, it was a series of articles on organic remains in the *Penny Magazine* – a joint venture of the Society for the Diffusion of Useful Knowledge and the publisher Charles Knight, which had begun publication in only the previous year – that had convinced Susan of the interest of geological study. Her brother's so-called 'Monster M' would itself soon become a regular feature of such self-consciously popular publications that, as Susan put it, 'contain every kind of knowledge written so plainly with prints [...] [and] which the most foolish person can understand'.[5] That 'extraordinary quadruped, the megatherium', as the *Penny Magazine* excitedly informed its one hundred and sixty thousand or so readers, 'resembled an elephant, but one of a gigantic size [...] and it was covered with a coat of mail something like that of an armadillo'.[6] In the popular print culture that emerged in the 1830s, a similar sense of the pleasure and curiosity of geology that Darwin felt so forcefully in South America was conveyed far beyond just his family circle.

***

Despite his rather rapid attribution of the fossil bones as megatheroid, based largely on the fragments of tessellated carapace found with them that Cuvier had earlier identified as a defining characteristic of the genus, Darwin's interest really lay in describing the geological locations in which the remains were found. He was especially concerned with the marine shells, largely identical with living species, that helped date the remains as relatively recent geologically.[7] Indeed, when discussing the bones with John Henslow in March 1834, he confessed his 'entire ignorance of comparative Anatomy', while six years later, and much more publicly, he acknowledged at the beginning of the portion of the *Zoology of the Voyage of HMS Beagle* devoted to 'Fossil Mammalia' (1840) that he did 'not possess the knowledge requisite for such an undertaking' and accordingly had devolved the task to another naturalist who was better qualified.[8] The subsequent appearance of the fossil remains in a range of popular publications, from demotic journalism to some of the best-known novels of the period, was therefore not mediated by their original discoverer but rather by the more anatomically accomplished individual to whom the bones, now cleaned and divested of the gravel in which they were found, were sent. This is particularly significant because in 1836, Darwin, recently returned to London, chose to present most of his fossil specimens to the Hunterian Museum at the Royal College of Surgeons. Here they were examined by the Hunterian's rising star of comparative anatomy, Richard Owen. Owen in fact also advised Darwin on where to deposit the remainder of his South American collections, and at this period the two men appear to have enjoyed a genuine and mutually beneficial friendship.

Owen, though, is today chiefly remembered as one of the most savage critics of the theory of evolution by natural selection, and his vocal opposition to Darwinism, and especially to his particular *bête noire*, Thomas Henry Huxley, has ensured that he has either been entirely written out of those triumphalist narratives of scientific progress predicated on evolution or else cast as the malevolent enemy of everything that is enlightened and secular. Janet Browne has memorably described Owen as the 'skeleton in the cupboard of evolutionary science', and it is important to recognize that he was by no means opposed to all forms of evolution.[9] In fact, from the mid-1840s onwards Owen became increasingly interested in a progressive process of transmutation that was attributable to what he called the 'axiom of the continuous operation of the ordained becoming of living things'.[10] He was nevertheless careful not to publish any views that would have been anathema to his scientific patrons amongst the political and intellectual establishment. Thus when Darwin broke his own silence on species transmutation with the publication of *On the Origin of Species* in 1859, that book's presentation of Owen 'as being firmly convinced of the immutability of species' was, as Darwin conceded ten years later, 'a preposterous error'.[11] Even while his reputation as an arch opponent of all forms of evolution is largely erroneous, Owen's genuine antagonism towards Darwin and his acolytes, expressed most clearly in a brutal review of the *Origin* for the *Edinburgh Review* in 1860, certainly adds a further relevance, and perhaps even an irony, to what happened to Darwin's *Beagle* fossils, following Owen's identification and elaboration of them, in nineteenth-century popular culture. At the same time, drawing attention to the rather limited role played by Darwin in the cultural afterlife of his *Beagle* fossils, and emphasizing the leading role taken by the customarily neglected Owen, provides an important counterexample to what Paul White has termed the 'Darwin-o-centrism' of much recent scholarship on nineteenth-century science and culture.[12]

In the opening words of the *Origin of Species*, Darwin reflected that

> When on board H.M.S. 'Beagle', as naturalist, I was much struck with certain facts in the distribution of the inhabitants of South America, and in the geological relations of the present to the past inhabitants of that continent. These facts seemed to me to throw some light on the origin of species—that mystery of mysteries [...] On my return home, it occurred to me [...] that something might perhaps be made out on this question.[13]

But while Darwin himself was privately speculating on the evolutionary implications of the apparent affinities between the megatherium and the present-day inhabitants of South America, it was Owen's own interpretation of the relation between the creature's seemingly ponderous anatomy and its peculiar feeding habits, based on Darwin's remains but invested with an implicit natural theological meaning, that was taken up in a variety of public forums. In interpreting the remains of the megatherium brought back by Darwin, Owen deployed the technique of functional correlation, a method of palaeontological reconstruction in which each element, or part, of an animal is presumed to correspond mutually with all the others, so that a carnivorous tooth must be accompanied by a particular kind of jawbone that facilitates the consumption of flesh, and so on; thus any part, even the mere fragment of a bone, necessarily indicates the configuration of the integrated whole. This principle, which proposed that animal structures were shaped to their adaptational needs, or in other words that form was determined by function, had been developed by Cuvier in the last decade of the eighteenth century, but it had become increasingly central to the English tradition of natural theology, as it seemed to show that only providential design could have produced such perfectly integrated mechanisms as the megatherium.

It was, Charles Carter Blake wrote in the *Geologist* in 1862, the 'labours of Mr. Charles Darwin' that 'made the form [...] of the Megatherium [...] familiar to us', and he noted that the 'past forms' of this creature 'which the acumen and critical skill of Professor Owen, or the sagacity and hardy research of Mr. Darwin have unveiled to us' were now 'enshrined in our Museums, or by their restorations which ornament our Crystal Palace'.[14] This last reference was to the spectacular glass and iron structure which had housed the Great Exhibition in 1851 and been rebuilt three years later on Sydenham Hill in South London, with its grounds full of visual delights such as the life-sized concrete models of reconstructed prehistoric creatures designed by Benjamin Waterhouse Hawkins. Although a protégé of Owen's, Blake gave Darwin equal billing with his mentor when observing that 'South American Palaeontology may well be proud that such labourers as these exist to illustrate its phases, or to demonstrate its significance'.[15] Yet in museums and popular sites, as well as in journalism and novels, it was almost exclusively Owen's functionalist account of the remains that prevailed. Even the megatherium's popular association with the sin of slothfulness, on account of its taxonomic affinities with the putatively lethargic sloths of present-day South America, was, as Alan Rauch has argued, inflected by Owen's influential description of the creature's habits.[16] In an era of enormous social, technological and cultural change, moreover, the lumbering but seemingly perfectly adapted megatherium, reconstructed from tiny fragmentary parts, offered ways of understanding novel technologies such as railway locomotives or new publishing forms such as the lengthy novels read in small serial parts, which ensured that Owen's functionalist interpretation of Darwin's *Beagle* fossils would continue to circulate in popular culture for much of the nineteenth century.

Owen's taxonomic assignment of the *Beagle* fossils at the Hunterian modified several of the attributions Darwin had made in the field, especially with regard to the megatherium, which he showed was not covered in armour, as Darwin, following Cuvier, had assumed. In fact, Owen showed that the tessellated hide that had so interested Darwin's sisters belonged to another mammal altogether, which he termed the glyptodon. That Darwin's error has hitherto been rarely acknowledged – with Sandra Herbert, in 2005, describing the 'false interpretation' as a 'curious and unknown story' – gives an indication of the distorting effect of the scholarly 'Darwin-o-centrism' mentioned earlier in representing him as an exceptional and uniquely brilliant scientific and cultural figurehead.[17] By separating the megatherium from the armadillo-like glyptodon, Owen augmented the case for its much closer relation to the sloth, a connection that was also strengthened by another new megatheroid characteristic gleaned from Darwin's specimens, the presence of one further upper molar than had been discerned by Cuvier and other earlier anatomists. This previously missing fifth molar revealed the true structure of the megatherium's dentition and allowed Owen to demonstrate that its staple diet, mostly consisting of leaves and soft sprouts, again corresponded with that of the sloth rather than the armadillo. Importantly, accommodating this sloth-like dentition with the clearly fossorial and therefore non-sloth-like character of the megatherium's enormous claws, established by previous remains and now confirmed by Darwin's specimens, enabled Owen to infer much about the overall structure of the creature as well as the nature of its mysterious and much contested feeding habits. As he wrote in the *Zoology of the Voyage of HMS Beagle*: 'In the remains of the Megatherium we have evidence of the frame-work of a quadruped equal to the task of undermining and hawling [*sic*] down the largest members of a tropical forest [...] which gives the explanation of the anomalous development of the pelvis, tail, and hinder extremities [which allowed it to remain upright]. No wonder [...] that their type of structure is so peculiar; for where shall we now find quadrupeds equal, like them, to the habitual task of uprooting trees for food?'[18]

Starting from the single extra tooth in the fragmentary cranium brought back by Darwin, Owen was able to explain the relation between all the apparently anomalous elements of the megatherium's anatomy and to show that their harmonious relation to each other allowed a mode of feeding that, while ungainly, was closely suited to the particular environment in which the gigantic creature had lived. Owen's account of the perfect functional correspondence between the ostensibly ill-proportioned megatherium's peculiar feeding habits and its complicated anatomical structure politely amended William Buckland's earlier explanation, in his Bridgewater Treatise *Geology and Mineralogy Considered with Reference to Natural Theology* (1836), of how its 'egregious apparent monstrosity' and the seeming 'incongruities of all its parts' were 'in reality systems of wise and well contrived adaptation'.[19] With the correction of their original discoverer's inexpert assignments, Darwin's fossil remains afforded the means for an updated natural theological reading of the most celebrated of the prehistoric megafauna, in which, as Owen wrote later in his *Memoir on the Megatherium* (1861), 'the fertility of the Creative resources is well displayed'.[20] It should nevertheless be noted that, as Nicolaas A. Rupke has argued, Owen's nods to natural theology were generally a means of maintaining the support of the establishment Oxbridge clique around Buckland, and he himself remained more concerned with the secondary laws by which the deity worked (which, as was suggested above, could be accommodated with a teleological process of evolution) as well as later adopting an alternative archetypal understanding of vertebrate design.[21]

***

Along with this underlying natural theological agenda, Owen's skilful deduction of the harmony of the megatherium's anatomy and habits, which went on to show that its heavily armoured skull was adapted to allow it to pull down large trees without the attendant danger of concussion, also invested the fossil bones with some of the pleasure and curiosity that Darwin and his sisters had initially felt upon their discovery. Readers of Owen's *Description of the Skeleton of an Extinct Gigantic Sloth* (1842) were instructed,

> now let us picture to ourselves the massive frame of the Megatherium, convulsed with the mighty wrestling, every vibrating fibre reacting upon its bony attachment

with a force which the sharp and strong crests and apophyses loudly bespeak:– extraordinary must have been the strength and proportions of that tree, which rocked to and fro, to right and left, in such an embrace, could long withstand the efforts of its ponderous assailant.[22]

Eight years later this invitation was taken up by the clergyman and novelist Charles Kingsley, whose *Alton Locke* (1850) features a dream sequence in which the eponymous tailor and poet imagines himself 'a mylodon among South American forests' and describes how he would

> plant my hinder claws at some tree-foot [...] and clasp my huge arms round the stem of some palm or tree-fern; and then slowly bring my enormous weight and muscle to bear upon it, till the stem bent like a withe, and the laced bark cracked, and the fibres groaned and shrieked, and the roots sprung up out of the soil; and then, with a slow circular wrench, the whole tree was twisted bodily out of the ground.[23]

The *Mylodon* was a subgenus of the *Megatherium* genus, which Owen had first identified from Darwin's *Beagle* fossils and, in recognition of the fact, had given the full taxonomic name *Mylodon darwinii*. While the language of this passage from *Alton Locke* is borrowed directly from Owen – the Mylodon even notes that he has 'fractured my own skull three or four times' – Kingsley places it in a developmental sequence similar to that of the recent anonymous evolutionary bestseller *Vestiges of the Natural History of Creation* (1844), so that Locke progresses abruptly from his megatheroid condition to become a baby-ape and then finally a primitive human.[24] Such transmutationist implications were, very broadly, similar to Darwin's own private speculations about the relation of type between the extinct megatherium and the present-day armadillo, which, curiously, he maintained into the mid-1840s despite Owen's removal of the apparently linking carapace and establishment of a much better case for the relation between glyptodon and armadillo.[25] It is striking, however, that nowhere else in the popular response to the megatherium would these proto-evolutionary inferences be raised. Instead the central focus would be on the harmony of its oversized but perfectly correlated parts as established by Owen.

Before returning to other novelistic representations of the megatherium, I want to focus on its relation to a technological innovation that was almost exactly contemporaneous with Owen's examination of Darwin's South American fossils and which would have a transformative impact on the country to which the scientific traveller returned in the mid-1830s. In 1837, as Owen worked on the megatherium remains at the Hunterian, Robert Stephenson and Edward Bury were both completing their designs for rival railway locomotives which heralded the beginning of the great Victorian railway boom (the source, of course, of many domestic fossil finds in the earthworks necessary to lay down tracks).[26] Nor was it long before these slightly fearful marvels of engineering, which had opened up even the furthest reaches of the country, were being compared to the prehistoric creature whose anatomy and habits had been revealed by Owen. An article from *Fraser's Magazine* in 1846, for instance, advised prospective travellers to northern Scotland: 'As you near Arbroath, probably your eye may catch something skimming rapidly along the beach, like [...] a megatherium smoking a cigar. It is a train on the Dundee and Arbroath railway.'[27] The rather louche masculine pleasures of smoking tobacco were not among the creature's habits inferred by Owen, although his wife Caroline did often encourage him to 'smoke cigars all over the house' to counteract the smell of putrid animal cadavers in their cramped apartment above the Hunterian Museum.[28] Nevertheless, the strikingly discordant analogy employed in *Fraser's* highland travelogue not only reinvested the megatheroid bulky frame with some of the amusement and pleasure felt by Darwin and his sisters about the curious 'Monster M' but, more importantly, located its articulated carcase as the equivalent of the locomotive's steam-powered engine with the upwardly protruding cigar as its funnel.

A decade later, when Benjamin Waterhouse Hawkins's reconstruction of a megatherium had taken its place, with other concrete models of antediluvian megafauna, in the grounds of the Crystal Palace at Sydenham, a writer in the *London Quarterly Review* described the view from the roof of the Palace in similar terms: 'At your feet is the park, on the furthest edge of which the geological monsters stand, while immediately beyond them comes, in full career, as if mocking their impotence, the *megatherium* of the nineteenth century, blowing like a whale, snorting like a wild horse, and making the welkin resound with the thunder of his train'.[29] Several commentators on the Crystal Palace noted that, when arriving by the special train service from London Bridge, the first thing to be seen of the grounds from the window of the railway carriages was the rather disconcerting prehistoric reconstructions, and for many visitors there was clearly an even closer link between their own mode of transport and the model megatherium, who was figured, in accordance with Owen's account, clasping a tree trunk in its forearms.

But if the railway locomotive was the megatherium of the nineteenth century, it was not just because of their shared fearful bulk or potentially destructive power (and there were many anxieties in the 1840s that the continually spreading railway was tearing the country up as the megatherium did its tropical forest). Rather, they were also both instances of complex but perfectly integrated mechanisms that, as in the classic instance of the watch from William Paley's 1802 work *Natural Theology*, necessarily indicated the presence of a designer, whether human or divine. Even before the advent of the railways, Paley himself had pointed to 'steam-engines' which 'deriv[e] their curious structures from the thought and design of their inventors' as a further analogy of providential design.[30] Buckland had compared the anatomy of the megatherium, in his Bridgewater Treatise, to the 'hammer and anvil of an anchorsmith', which 'though massive' were 'neither clumsy nor imperfect', while Owen's amended and less overtly natural theological account based on Darwin's fossils continued to make similar references to the 'efficiency of [its] masticating machinery'.[31] The very model of the megatherium witnessed through the train window at the Crystal Palace was itself constructed on similarly Paleyan mechanistic principles, for its creator, Waterhouse Hawkins, later reflected that the artist who wished to represent such a

> living mechanism [...] can never fully succeed without having carefully studied the parts of its machinery [...] Neglecting this, he will be in the position of the engineer who tries to understand or represent some complex machine, of whose structure and uses he knows no more than an outside glance has told him.[32]

The megatherium, even in the very same reconstructions which Blake had identified as being modelled directly upon Darwin's original fossil discoveries, was associated with an implicitly natural theological understanding of its structure, based on its apparent resemblance to railway locomotives and other items of machinery, that, in line with Paley, had been inaugurated by Buckland and then brought to wider attention by Owen. And this mechanistic interpretation of the creature was still current at least as late as the 1870s, with a female railway passenger in Benjamin Disraeli's novel *Lothair* (1870) describing the 'whirl and whistling, and the wild panting of the loosened megatheria who drag us'.[33] Ironically, representations of the megatherium in various facets of Victorian popular culture therefore upheld precisely the Paleyan interpretation of its anatomy that the original discoverer of vital parts of its remains was, first privately but now openly, attempting to undermine.

***

The regular appearance of the megatherium in Victorian fiction is hardly surprising given the intense curiosity provoked by the creature's sheer size and enigmatic style of life, but many of these novelistic representations also had similarly natural theological connotations that derived from Owen's interpretation of Darwin's *Beagle* fossils. The extent to which Owen was identified, and sometimes actually conflated, with the peculiar sloth-like quadruped is evident from a briefly fashionable but now long forgotten contribution to the genre of sensation fiction, *Blount Tempest* (1865) by John Chippendale Montesquieu Bellew, in which the lawyer at the centre of a convoluted plot of murder and hidden identity briefly recalls attending – as did Bellew himself – the same Lancastrian school as the celebrated 'Professor Megatherium Bowen'.[34] Many writers of fiction also used the creature's complex anatomy as a model for their own works, and especially for the newly fashionable genre of enormously long novels published in monthly instalments.[35] William Makepeace Thackeray, the famous author of *Vanity Fair* (1847), for instance, used the 'Megatherium Club' as a habitual nickname for the Athenaeum Club in London from his bohemian days at *Punch* in the mid-1840s and continued to do so in many of his best-known novels.[36] While this droll epithet presumably referred to the sheer size of the famous establishment in Pall Mall, as well as to the Graeco-Roman etymology of its convoluted name, it almost certainly originated with the celebrated elaboration of the megatherium's structure and habits made by Owen, a Club member like Thackeray, only a few years earlier. The comic potential of the purportedly unwieldy creature was also used by Thackeray in a number of other contexts, such as the exotic dish of 'Cotelettes à la Megatherium' served to the pretentious gourmands of London high society.[37]

Thackeray's very first reference to the prehistoric giant which seems to have captivated his imagination came in his Christmas book for 1846, *Mrs. Perkins's Ball*, and, significantly, he initially perceived the megatherium's immense and cumbersome frame as analogous not to a large metropolitan club but rather to a ruinously colossal literary publication. Amongst the characters introduced in *Mrs. Perkins's Ball* is 'Poseidon Hicks, the great poet', who is the author of, amongst other works, 'The Megatheria'. Although this epic poem is '"a magnificent contribution to our pre-adamite literature", according to the [...] reviews', Thackeray's more sceptical narrator, Michael Angelo Titmarsh, reflects: 'I know that poor Jingle, the publisher, always attributed his insolvency to the latter epic, which was magnificently printed in elephant folio'.[38] This was only the first of many such disparaging references to huge and ungainly literary megatheriums throughout the nineteenth century, with American 'paper-makers' even 'coming out with a "megatherium"' sheet size that would have dwarfed Jingle's elephant folio.[39] Within eight years, however, Thackeray himself would exhibit a very different, and considerably more subtle and sympathetic understanding of the putative relation between the megatherium and amply proportioned works of literature.

Whereas in *Mrs. Perkins's Ball*, Titmarsh's ironic narration indicates that 'The Megatheria' is merely a verbose and grandiloquent epic whose unwieldy size has bankrupted its publisher, the narrator of Thackeray's serial novel

The Newcomes, Arthur Pendennis, actually likens his own novelistic effusions to the same enormous creature. Pendennis reflects that,

> As Professor Owen [...] takes a fragment of a bone, and builds an enormous forgotten monster out of it, wallowing in primaeval quagmires, tearing down leaves and branches of plants that flourished thousands of years ago, and perhaps may be coal by this time – so the novelist puts this and that together: from the footprint finds the foot; from the foot, the brute who trod on it; from the brute, the plant he browsed on, the marsh in which he swam – and thus in his humble way a physiologist too, depicts the habits, size, appearance of the beings whereof he has to treat; – traces this slimy reptile through the mud, and describes his habits filthy and rapacious; [...] points out the singular structure of yonder more important animal, the megatherium of his history.[40]

The wrappers of *The Newcomes*' monthly numbers had already carried advertisements for Owen's more popular publications in which such functionalist methods of reconstruction were regularly adumbrated, and now the very same techniques were incorporated within the novel's own fictional frame as an analogy for its style of narration.[41] However, the passage seems, at first, merely to reflect Thackeray's characteristic cynicism about narratorial omniscience, with Pendennis appearing to suggest that much of the details of the history of the most respectable Newcome family are based on questionable inferences and dubious hypothetical reconstructions. This is how the passage has usually been interpreted by critics of *The Newcomes*, with George Levine contending that the palaeontological method is invoked 'half-mockingly'.[42] It is, however, important to balance the recognition of Thackeray's habitual mordant cynicism with the knowledge that he might have gained a firsthand – if slightly misremembered – knowledge of such palaeontological procedures from Owen himself.

From the early 1840s, Thackeray and Owen would have regularly encountered each other in the familiar purlieus of the metropolitan literary and intellectual elite, including the celebrated Pall Mall establishment that Thackeray had dubbed the Megatherium Club. Certainly, Owen recorded – in the first extant evidence of their acquaintance – that at a dinner at the Royal Academy on 4 May 1850, he saw 'Thackeray, who sent to me across the table to take a glass of wine'.[43] Owen reflected that with such regular social encounters amongst the intelligentsia, 'London is the place [...] for interchange of thought', and while Thackeray appears not to have read Owen's work, with none of his publications featuring in the novelist's library at the time of his death, their friendship would undoubtedly have brought him into contact with aspects of Owen's scientific thought.[44] Thackeray, for instance, was a fellow guest at a dinner at the home of Lord Ashburton on 11 May 1855 when Owen discoursed on palaeontology with the Duc d'Aumale, who, as Caroline Owen recorded, 'evidently knows something of fossils'.[45] The evidence of the precise nature of their friendship remains patchy, but Thackeray's acquaintance with Owen, who was, of course, renowned for his account of the perfectly integrated design of the ostensibly ill-proportioned megatherium, certainly coincides with a conspicuous shift in his treatment of the sloth-like quadruped as a model for the formal structure of literary works and suggests that the invocation of the functionalist methods that had made Owen famous was not predominantly cynical or sardonic.

Significantly, Pendennis's portrayal of the 'novelist [who] puts this and that together' as, in the same vein as Owen, 'a physiologist too', becomes still more pertinent in relation to the process of serialization in which Thackeray was engaged when writing *The Newcomes*. Like the palaeontologist using Cuvierian functionalist methods, or Owen working on the fragmentary remains brought back by Darwin, the serial novelist, who regularly wrote the next instalment only after the previous one was already published, had painstakingly to relate each individual part to a larger and often still conjectural narrative whole in order to build up both character and plot. Meanwhile, the novel's expectant audience, whose practice of reading similarly involved moving from part to whole, were left to predict how the events of each number would fit in – or correlate – with that overall framework. Owen seems to have appreciated that his famed elaboration of the paradoxically clumsy yet perfect giant South American ground-sloth might have a particular pertinence for novelists, sending a 'gift' of his *Memoir on the Megatherium* to George Eliot in January 1861 'as a sign of the pleasure [he] had had in "The Mill on the Floss"'.[46] It was, the book's recipient responded modestly, 'a very good and graceful thing for the greater worker to help the less in this way', and incidental allusions to the 'gouty humours of Lord Megatherium' or taking 'an antediluvian point of view' so as not to 'do injustice to the megatherium' appeared in both *Middlemarch* (1871–2) and *Daniel Deronda* (1876), although, notably, not in relation to the structure of these serialized novels.[47]

Henry James famously derided *The Newcomes* as a 'large loose baggy monster' (a description that has a lot of parallels with earlier accounts of the awkward and ungainly megatherium, or 'Monster M' as Susan Darwin called it), and there were many fears amongst other critics that serialized novels, built up incrementally from often disparate parts and lacking the formal coherence and design characteristic of a discrete monograph, would become unwieldy and absurdly incongruous monsters.[48] But if, as Thackeray states, the serialized novel assumes the 'singular structure' of that 'important animal, the megatherium of his [i.e. the novelist's] history', it is not simply on account of its prohibitive dimensions, as with Poseidon Hicks's gargantuan epic poem. Rather, Owen, as the allusion to him in *The Newcomes* suggested, had been able, beginning with only the single extra tooth in a fragmentary cranium brought back from South America by Darwin, to explain the necessary connection between all the apparently anomalous elements of the megatherium's anatomy and show their harmonious relation to one another.

In his *Description of the Skeleton on an Extinct Gigantic Sloth* (1842) Owen observed that the principle manifested in the 'admirable adaptation' of the multifaceted 'fore-foot of the extinct Megatheroid quadrupeds' was 'beautifully set forth by the poet' in the following lines:

> In human works, though labour'd on with pain,
> A thousand movements scarce one purpose gain:
> In God's, one single can its end produce;
> Yet serve to second too some other use.[49]

When properly understood, the apparent monstrosity and incongruity of the megatheroid structure, Owen suggested, in fact corresponded with the neoclassical formal coherence of Alexander Pope's *Essay on Man* (1732–4). Pope's epigrammatic poem was an exemplar of Enlightenment natural theology, especially that adumbrated by Lord Bolingbroke, and Owen's use of it to depict the exquisite functional adaptations of a creature in which, as he later observed, the 'fertility of the Creative resources is well displayed', implies a connection between aesthetic and divine design that accords with his distinctly Keatsian insistence that the 'laws of correlation rightly discerned [...] are [...] as beautiful as they are true'.[50] As a voracious and miscellaneous reader of literature, Owen saw no difficulties in switching from Enlightenment to Romantic analogies.[51] The serialized novel might too, as a species of literary megatherium, reveal an underlying design behind its seemingly ill-proportioned parts that would render it as aesthetically unassailable as the most revered literary works of the previous century. Thackeray's comparison of the loose, baggy organization of *The Newcomes* to the 'singular structure' of the megatherium therefore suggested a parallel between the initially enigmatic but nonetheless perfectly integrated designs of the serial novelist and the omnipresent author of the natural world, that once more reinforced the natural theological interpretation of the megatherium in nineteenth-century popular culture.

***

The central argument of this chapter has been that once the fossil remains unearthed by Darwin during the South American leg of the *Beagle* voyage were taken possession of by the Hunterian Museum in London, they soon assumed the range of meanings attributed to them in Owen's functionalist interpretation of the megatherium's structure and habits. While there were still some acknowledgements of Darwin's role in finding and making the remains available, as with Blake's comments in the *Geologist* in 1862, representations of the megatherium in a variety of popular cultural forms, from journalism and life-size reconstructions to serial novels, almost invariably drew on Owen's particular account of the creature and, more significantly, often retained the underlying natural theological meanings that Owen, following Buckland, had found in the perfectly designed mechanisms of its anatomical structure. In his autobiography, Darwin, reflecting on his thoughts 'in relation to the transmutation of species', recalled that 'I had been deeply impressed by discovering in the Pampean formation great fossil animals covered with armour like that on the existing armadillos.'[52] These were, of course, the combined remains of megatheriums and glyptodons that Darwin had wrongly, although in accordance with Cuvier, conflated into a single creature. But the megatheroid structure elaborated by Owen from the same fossils, now divested of its erroneous armour, would, as well as becoming an icon of natural theological interpretation in Victorian popular culture, also emerge, following the publication of *On the Origin of Species* in 1859, as a bulwark against Darwinian evolution.

In the conclusion to his *Memoir on the Megatherium* from 1861, Owen insisted that, despite initial appearances, modern South American animals were 'specifically distinct' from the megatherium, and by such 'well marked [...] characters [...] as no known outward influences have been observed to produce by progressive alteration of structure'. Owen was careful not to mention Darwin's recently published work on the same subject and instead referred his remarks to the refutation of what he called 'LAMARCK's progressive hypothesis of the origin of species by transmutation', but most readers in the early 1860s would doubtless have also made the connection with the *Origin*.[53] The only mention of the now highly topical Darwin in Owen's memoir was actually at the beginning, when Owen referred to what he had found out about the 'affinities of the Megatherium' from the 'portions of teeth, obtained by MR. CHARLES DARWIN at Punta Alta in Northern Patagonia'.[54] For Owen, less than a year after his infamous attack on the *Origin* in the *Edinburgh Review*, there was surely an irony in recording his debt to Darwin at the beginning of a work which closed by deploying the very osseous remains discovered by Darwin in South America to impugn, at least implicitly, Darwin's own evolutionary theories.

Notes

1 *The Correspondence of Charles Darwin*, ed. Frederick H. Burkhardt et al., 18 vols (Cambridge: Cambridge University Press, 1983–), 1:379.
2 Ibid., 1:276.
3 Robert FitzRoy, *Narrative of the Surveying Voyages of His Majesty's Ships 'Adventure' and 'Beagle' between the Years 1826 and 1836*, 3 vols (London: Henry Colburn, 1839), 2:107.
4 *Correspondence of Charles Darwin*, 1:299.
5 Ibid.
6 'Organic Remains', *Penny Magazine* 2 (1833): 395. For the circulation figures of the *Penny Magazine* in the early 1830s, see Scott Bennett, 'Revolutions in Thought: Serial Publication and the Mass Market for Reading', in *The Victorian Periodical Press: Samplings and Soundings*, ed. Joanne Shattock and Michael Wolff (Leicester: Leicester University Press, 1982), 236.
7 See Sandra Herbert, *Charles Darwin, Geologist* (Ithaca: Cornell University Press, 2005), 163 and passim.
8 *Correspondence of Charles Darwin*, 1:368; Charles Darwin, 'Preface', in *Zoology of the Voyage of HMS Beagle*, ed. Charles Darwin, 5 vols (London: Smith, Elder, 1839–43), 1:iii.
9 Janet Browne, 'Natural Causes', *Times Literary Supplement* (12 August 1994): 3.
10 Richard Owen, *Palaeontology*, 2nd ed. (Edinburgh: Adam and Charles Black, 1861), 3.
11 Charles Darwin, *On the Origin of Species*, 5th ed. (London: John Murray, 1869), xx.
12 Paul White, 'Science, Literature, and the Darwin Legacy', *19: Interdisciplinary Studies in the Long Nineteenth-Century* 11 (2010): 3, http://www.19.bbk.ac.uk/index.php/19/issue/view/79 (accessed 29 July 2012).
13 Charles Darwin, *On the Origin of Species* (London: John Murray, 1859), 1.
14 Charles Carter Blake, 'Past Life in South America', *Geologist* 5 (1862): 329.
15 Ibid.
16 Alan Rauch, 'The Sins of Sloths: The Moral Status of Fossil Megatheria in Victorian Culture', in *Victorian Animal Dreams: Representations of Animals in Victorian Literature and Culture*, ed. Deborah Denenholz Morse and Martin A. Danahay (Aldershot: Ashgate, 2007), 221–2.
17 Herbert, *Darwin, Geologist*, 303–4. See also Jim Endersby, 'Escaping Darwin's Shadow', *Journal of the History of Biology* 36 (2003): 385–403; and Gowan Dawson, 'Science and Its Popularization', in *The Cambridge Companion to English Literature 1830–1914*, ed. Joanne Shattock (Cambridge: Cambridge University Press, 2010), 172–4.
18 Richard Owen, 'Fossil Mammalia', in *Zoology*, 1:106.
19 William Buckland, *Geology and Mineralogy Considered with Reference to Natural Theology*, 2 vols (London: William Pickering, 1836), 1:144–5.
20 Richard Owen, *Memoir on the Megatherium* (London: Williams and Norgate, 1861), 41.
21 Nicolaas A. Rupke, *Richard Owen: Victorian Naturalist* (New Haven: Yale University Press, 1994), 152–6 and passim.
22 Richard Owen, *Description of the Skeleton of an Extinct Gigantic Sloth* (London: John Van Voorst, 1842), 148.
23 Charles Kingsley, *Alton Locke, Tailor and Poet*, 2 vols (London: Chapman and Hall, 1850), 2:217–18.
24 Ibid., 2:219. Kingsley, of course, later responded enthusiastically to Darwin's *Origin*, accommodating its evolutionism within his own Anglican natural theology. See Bernard Lightman, *Victorian Popularizers of Science: Designing Nature for New Audiences* (Chicago: University of Chicago Press, 2007), 71–81.

25 See Herbert, *Darwin, Geologist*, 306–7.
26 See Michael J. Freeman, *Railways and the Victorian Imagination* (New Haven: Yale University Press, 1999).
27 'Manners, Traditions and Superstitions of the Shetlanders', *Fraser's Magazine* 33 (1846): 636.
28 Richard S. Owen, *The Life of Richard Owen*, 2 vols (London: John Murray, 1894), 1:296.
29 'The Crystal Palace', *London Quarterly Review* 3 (1854): 235.
30 William Paley, *Natural Theology* (London: R. Faulder, 1802), 71.
31 Buckland, *Geology and Mineralogy*, 1:159; Richard Owen, *Memoir on the Megatherium*, 35.
32 Benjamin Waterhouse Hawkins, *The Artistic Anatomy of the Dog and Deer* (London: Winsor and Newton, 1876), 18.
33 Benjamin Disraeli, *Lothair*, 3 vols (London: Longmans, Green, 1870), 1:108.
34 J. C. M. Bellew, *Blount Tempest*, 3 vols (London: Hurst and Blackett, 1865), 3:114.
35 This argument is developed in more detail in Gowan Dawson, 'Literary Megatheriums and Loose Baggy Monsters: Paleontology and the Victorian Novel', *Victorian Studies* 53 (2011): 203–30.
36 See, for instance, [W. M. Thackeray], 'A Little Dinner at Timmins's', *Punch* 15 (1848): 5; and W. M. Thackeray, *The History of Pendennis*, 2 vols (London: Bradbury and Evans, 1849), 1:282.
37 [Thackeray], 'A Little Dinner at Timmins's', 5.
38 [W. M. Thackeray], *Mrs. Perkins's Ball* (London: Chapman and Hall, 1846), 19.
39 Edward P. Hingston, *The Genial Showman. Being Reminiscences of the Life of Artemus Ward*, 2 vols (London: John Camden Hotten, 1870), 1:191.
40 W. M. Thackeray, *The Newcomes*, 2 vols (London: Bradbury and Evans, 1854–55), 1:81.
41 See, for instance, *'The Newcomes' Advertiser* 9 (1854): 3.
42 George Levine, *The Realistic Imagination: English Fiction from 'Frankenstein' to 'Lady Chatterley'* (Chicago: University of Chicago Press, 1981), 164.
43 Richard S. Owen, *Life*, 1:358.
44 Ibid., 2:23; see J. H. Stonehouse, *Catalogue of the Library of Charles Dickens and W. M. Thackeray* (London: Piccadilly Fountain Press, 1935).
45 Richard S. Owen, *Life*, 2:6.
46 *The George Eliot Letters*, ed. Gordon S. Haight, 9 vols (New Haven: Yale University Press, 1954–78), 3:373.
47 Ibid.; George Eliot, *Middlemarch*, 4 vols (London: William Blackwood, 1871–72), 1:98; George Eliot, *Daniel Deronda*, 4 vols (London: William Blackwood, 1876), 4:50.
48 Henry James, *The Tragic Muse*, 2 vols, New York ed. (New York: Charles Scribner, 1908), 1:x.
49 Richard Owen, *Skeleton of an Extinct Gigantic Sloth*, 149.
50 Richard Owen, *Monograph on the Fossil Mammalia of the Mesozoic Formations* (London: Palæontographical Society, 1871), 100.
51 On Owen's extensive literary reading, see Gowan Dawson, '"By a Comparison of Incidents and Dialogue": Richard Owen, Comparative Anatomy and Victorian Serial Fiction', *19: Interdisciplinary Studies in the Long Nineteenth-Century* 11 (2010). Online: http://www.19.bbk.ac.uk/index.php/19/issue/view/79 (accessed 29 July 2012).
52 *The Life and Letters of Charles Darwin*, ed. Francis Darwin, 3 vols (London: John Murray, 1887), 1:82.
53 Richard Owen, *Memoir on the Megatherium*, 81.
54 Ibid., 9.

Chapter 7

'NO SUCH THING AS A FLOWER [...] NO SUCH THING AS A MAN': JOHN RUSKIN'S RESPONSE TO DARWIN

Clive Wilmer

In the last of the five volumes of *Modern Painters*, published in 1860, 17 years after the project began, John Ruskin proclaims what he calls 'the Law of Help'. He has been talking about composition in painting – about the way the individual parts of a picture contribute to the whole – and he then goes on to affirm such collaboration as the ruling principle of nature itself:

> [I]n a plant, the taking away of any one part [...] injure[s] the rest. Hurt or remove any portion of the sap, bark, or pith, the rest is injured. If any part enters into a state in which it no more assists the rest, and has thus become 'helpless', we call it 'dead'.
>
> The power which causes the several portions of the plant to help each other, we call life. Much more is this so in an animal. (7:205)[1]

And of course (he goes on) still more so in humans. He goes so far as to retranslate the old Anglo-Saxon word 'holy' as 'helpful', so that God becomes 'the Helpful One' (7:206). This discussion completed, he then announces his 'Law':

> A pure or holy state of anything, therefore, is that in which all its parts are helpful or consistent. They may or may not be homogeneous. The highest or organic purities are composed of many elements in an entirely helpful way. The highest and first law of the universe – and the other name of life is, therefore, 'help'. The other name of death is 'separation'. Government and co-operation are in all things and eternally the Laws of Life. Anarchy and competition, eternally and in all things, the Laws of Death. (7:207)

The first readers of *Modern Painters* V seem not to have realized or anticipated that the volume was, in effect, Ruskin's signing off as primarily a writer on art. Six months later, he was to publish the first of the four essays on political economy he called *Unto This Last*. This new and unexpected book is both the fiercest and the most cogently argued of all Victorian attacks on free-market capitalism. In the third of the essays, Ruskin deliberately links his new book with the project of *Modern Painters* by quoting the very passage I have given: 'Government and co-operation are in all things the Laws of Life; Anarchy and competition the Laws of Death' (17:75)[2] – and he supports it in his conclusion with the dictum: 'THERE IS NO WEALTH BUT LIFE' (17:105).

The year 2010 saw the sesquicentennial anniversary of *Unto This Last*, serialized in 1860.[3] The social ethics of the book had been foreshadowed in Ruskin's earlier work, most notably in 'The Nature of Gothic', the central chapter of *The Stones of Venice* (1853). But this sudden shift from art and nature to economics and society – or perhaps I should say this tracing of the same law through all four categories – cannot have been unrelated to the most famous publication of the previous year: Charles Darwin's *The Origin of Species*; Or – if I may remind the reader of the subtitle – *The Preservation of Favoured Races in the Struggle for Life*, which first appeared in 1859. It is the subtitle that locates Darwin's book in the period of its publication, suggesting the necessity of capitalism to the formulation of its argument. It will be my purpose in this chapter to separate the title from the subtitle, so to speak, and to suggest that Ruskin's notorious aversion to Darwinism derives from an objection not so much to a theory of organic development as to the assumptions that made this particular theory possible and the assumptions it then inspired. That is partly to say that there was a social element in Ruskin's aversion: that it was more an objection to that element in Darwin that derived from the political economist T. R. Malthus, whose book *An Essay on the Principle of Population* (revised edition, 1803) had so radical an effect on Victorian society and social attitudes, than it was an objection to, say, Darwin's parallel debt to Sir Charles Lyell, the first volume of whose book *The Principles of Geology* (1930) had accompanied him on the voyage of the *Beagle*. (Ruskin's apparent admiration for Lyell, a uniformitarian rather than an evolutionist, emerges from his private correspondence. In his published writings he kept a politic silence.) At the same time, he could never have regarded the question of origins as purely a social matter. When he records the passion he felt as a boy of 15 on his first sight of the Alps, he notes that before the Romantic age, 'no child could have been born to care for mountains, or for the men who lived among them, in that way' (35:115). Any theory of the universe, for Ruskin, had to combine the love of humanity with the love of nature, both of them informed with the indivisible love of God.

Ruskin was not alone among Victorian intellectuals in having religious convictions that baulked at the theory of natural selection. Not only was he brought up as a scriptural Protestant, but he was taught at Oxford by William Buckland of *The Bridgewater Treatises* and became personally attached to him. Buckland, as one of Ruskin's biographers puts it, 'combined benevolent Christianity with an unparalleled scientific curiosity'.[4] For Ruskin as a writer on art, the whole point of painting lay in the tribute it paid to the loving work of God as the artist read it in nature: as he insisted, 'ALL GREAT ART IS PRAISE' (15:351). He believed as firmly as Buckland did that the beauties of nature were created by God for our pleasure and instruction, and much of *Modern Painters* is concerned with the relation of God's art to human art. The question that has to be asked, however, is how Ruskin understood natural beauty. Did he understand natural forms to be stable and unchanging? Was the world we live in now identical to the world that came into being when, in the words of the psalmist, God's 'hands prepared the dry land'?[5] Are such words as the psalmist's, moreover, to be taken literally? And even if they are, can they be understood in any way as the word of God? Ruskin provides no answer to many of these questions, but we do know that by the year of *The Origin of Species*, he had lost that Evangelical trust in the objective truth of the Bible, and his fascinated study of Alpine glaciers suggests that he knew perfectly well how the face of the earth had altered in the course of uncountable ages. In the previous year, he had been '*un*-converted' (29:89), as he tells us in *Fors Clavigera*, and had adopted 'the *Religion of Humanity*' (29:88n); and yet, remaining profoundly Christian in outlook, he was poised in a sort of spiritual limbo. His understanding of nature continued to include a concept of co-operation. As his eccentric dialogue on geology, *The Ethics of the Dust* (1866), suggests, he understood the relation of one mineral to an adjacent one as a matter of neighbourliness, and in *Modern Painters* V, the volume which includes 'the Law of Help', he writes of the leaves on a tree, first of all, as a society – he initially draws a comparison with bees – and then more specifically as a family:

> [E]very branch has others to meet or to cross, sharing with them, in various advantage, what shade, or sun, or rain is to be had. Hence every single leaf-cluster presents the general aspect of a little family, entirely at unity among themselves, but obliged to get their living by various shifts, concessions, and infringements of the family rules, in order not to invade the privileges of other people in their neighbourhood. (7:48)

I quote this almost at random from the section of the book entitled 'Of Leaf Beauty', which includes a closeness of botanical attention comparable to Darwin's and yet throughout its length draws on such social comparisons.

What emerges is a process of collaborative creation in nature, comparable to the account of medieval building he gives in 'The Nature of Gothic', of each individual self contributing to the greater whole.[6]

What is more, the conception of *human* creativity in 'The Nature of Gothic' is modelled on an evolutionary paradigm. God may have created a perfect world in the beginning, but human beings – as an effect of the Fall– are imperfect, and our imperfection, in Ruskin's view, is our glory. Imperfection, he says,

> is in some sort essential to all that we know of life. It is the sign of life in a mortal body, that is to say, of a state of progress and change. Nothing that lives is, or can be, rigidly perfect; part of it is decaying, part nascent. [...] All things are literally better, lovelier, and more beloved for the imperfections which have been divinely appointed, that the law of human life may be Effort, and the law of human judgment, Mercy. (10:203–4)

This explains the paradox that 'no architecture can be truly noble which is *not* imperfect' (10:202); being ourselves imperfect, we are enabled to reach out for perfection if never to attain it.[7]

This argument is never concerned exclusively with architecture. As he makes clear in the passage I have quoted and as his analogies with natural forms confirm, it is an argument about life itself. Moreover, certain of the plates in *The Stones of Venice* demonstrate how, in the practice of art and architecture, decorative forms evolve, the arrangement of the images on the page reminding one of diagrams in scientific textbooks.[8] Ruskin the geologist is at work here, as the very phrase 'the *stones* of Venice' suggests, but it is botanical diagrams that are called to mind in plate xx of volume 2, 'Leafage of the Venetian Capitals', where the subject is foliate growth (10:431). Plate xiv, 'The Orders of Venetian Arches', is indisputably evolutionary. In 37 diagrammatic images the plate shows how the simple, slightly elongated round arch of Byzantine origin develops into what Ruskin calls the Transitional style; how the Transitional, becoming increasingly sophisticated, evolves into the Gothic; and how the Gothic arch acquires the oriental profile so typical of Venice, gradually mutating into the refined and elaborate ogival arch of the fifteenth-century palaces (10: 290). For anyone interested in Darwin, the plate and others like it may call to mind the array of barnacles in plate 1, 'Balanus Tintinnabulum', of Darwin's *Monograph on the Sub-class Cirripedia* (1854),[9] published five years before *Origin*.

But to return to 'The Nature of Gothic', which predates Ruskin's 'unconversion' by five years: his insistence on human imperfection is inseparable from his conviction that humanity is *essentially* social, each individual a part of the larger body. This is not an argument for servile conformity; on the contrary, it emphasizes 'the individual value of every soul' (10:190), each artisan contributing from the uniqueness of his imagination to a great and glorious social artefact. That distinctiveness of the individual human is for Ruskin the grandest fact of life, but it is also characteristic of creation as a whole. The leaves on a single tree are recognizably leaves of the same kind but none of them is identical to the others and, far from being a cause of conflict, their distinctness is ground of their needful co-operation: their helpfulness. There is no doubt that Ruskin saw society as in effect an extended family; by implication, he also saw nature as a kind of society. This is perhaps to say that his theories of nature and society were as closely intertwined as Darwin's but that the social assumptions he made were strikingly different. Nevertheless, in his emphasis on unique and various development, Ruskin is essentially in agreement with Darwin. It was something that Darwin liked in Ruskin the man, delighting in 'the keenness of [his] observation and the variety of [his] scientific attainment' (36:553n). For strange though it may seem – given the vehemence of Ruskin's attacks on Darwin – the two men palpably enjoyed one another's company.

In a different intellectual climate Ruskin and Darwin might have become friends. They first met in 1837 when Darwin, just back from the voyage of the Beagle, read a paper to the Geological Society in London.[10] Darwin was 28, Ruskin, a student at Christ Church, Oxford, only 17 – extraordinarily young to be hobnobbing, as he was, with the likes of Lyell and Adam Sedgwick, but then Ruskin's first professional publications had been essays on Alpine geology contributed nearly three years earlier to the *Magazine of Natural History*. Unfortunately, we do not know what he made of Darwin's paper, but he was clearly pleased when, later that year, he met Darwin again at one of Buckland's celebrated 'breakfasts'. After the meal, we learn from one of Ruskin's letters, they 'got together and talked all evening' (36:14).

Thirty years passed before they met again, this time under the auspices of Ruskin's friend, Charles Eliot Norton, in 1868. By this time they were both famous, and Ruskin was on public record as an opponent of natural selection. They were nonetheless keen to meet and were to do so again on subsequent occasions. 'Ruskin's gracious courtesy', Norton reported, 'was matched by Darwin's charming and genial simplicity,' and he noticed how 'their animated talk afforded striking illustration of the many sympathies that underlay the divergence of their points of view, and of their methods of thought' (36:553n).

The fierceness of Ruskin's comments on Darwin and the warmth of their intercourse are surely related. His attacks are of a piece with those on the painter J. M. Whistler for courting abstraction, the atomization of nature (29:160), and on Charles Dickens for his obsession with urban death in *Bleak House* (34:271–3).[11] On meeting Ruskin in 1869, Henry James remarked that he seemed to have 'been scared back by the grim face of reality'.[12] I would prefer to call it the face of modernity. His problem with Darwinism was that, belonging to the same tradition as Darwin, he grasped its implications all too well. Darwin was pointing out a road that Ruskin had no wish to travel down. In the things he loved most – the flowers, the creatures, the clouds and the mountains – he could see nothing but the struggle for survival, the outcome of which could only be, as in *Bleak House*, senseless, degrading death.

Yet Ruskin understood perfectly well – and understood it before he lost his religious faith – that simple creationism would not do. He was acquainted with the geological literature that had helped to shape Darwin's theory. His famous statement of 1851 that he could hear the clinking hammers of the geologists 'at the end of every cadence of the Bible verses' (36:115) is sufficient evidence that he had grasped the implications of, for instance, Lyell's *Principles of Geology*, even if like Lyell he saw no need to follow the argument as far as a theory of evolution. He was also familiar with a range of materialist science – from Lamarck through Cuvier to Agassiz – which, though hardly to be understood as leading to natural selection, is nonetheless part of the intellectual atmosphere in which Darwin's theory was born. Even Buckland, though a committed catastrophist, had drawn attention to the difficulties of strict creationism. Though he had sought to reconcile modern geology with the biblical accounts of Creation and the Flood, he inadvertently and perhaps inevitably exposed the inherent problems and was forced to recognize that science and religion speak different languages. It was not just a matter of the seven days of Creation, easily understood as standing for seven eras, but, more troublingly, of such matters as the extinction of species. For these were evidence that creatures had died and even killed one another before the arrival of man and original sin, which (as Christian theology had always argued) 'Brought death into the world and all our woe'.[13] Ruskin's fiercely Evangelical mother – by no means an anti-intellectual, it should be said – was of the view that Buckland might have been wiser to 'let the Bible alone'.[14] We can see how Ruskin follows from Buckland in *Modern Painters* IV, the volume in which he discusses the 'materials' of Creation which the artist must learn to depict. There he seems to dismiss the sort of problem that Buckland had notably raised: 'What space of time was in reality occupied by the "day" of Genesis', he writes, 'is not, at present, of any importance for us to consider' (6:16), as if the question were one that had never troubled him. Earlier on in the same book, discussing 'The Firmament', he quotes from the Psalms: 'He bowed the heavens also, and came down; he made darkness pavilions round about him, dark waters, and thick clouds of the skies.'[15] Ruskin comments as follows:

> By accepting the words in their simple sense, we are thus led to apprehend the immediate presence of the Deity, and His purpose of manifesting Himself as near us whenever the storm-cloud stoops upon its course; while by our vague and inaccurate acceptance of the words we remove the idea of His presence far from us, into a region which we can neither see nor know; and gradually, from the close realisation of a living God who 'maketh the clouds his chariot' we refine and explain ourselves into a dim and distant suspicion of an inactive God, inhabiting inconceivable places, and fading into the multitudinous formalisms of the Laws of Nature. (6:110)

Elsewhere in the same volume he relishes such concretions of the psalmist's as the sentence quoted above: 'His hands prepared the dry land.'[16] Such statements would appear to belong to the same tradition of creationist or catastrophist argument as those pursued by Buckland. But are they in fact? It is clear that Ruskin repudiates the deistic implications of Paleyan natural theology – 'the multitudinous formalisms of the laws of Nature' – but is at the same time repelled by the tendency of liberal Christianity to symbolize the biblical narrative out of existence. It was the latter that finally put him out of sympathy with his ally in social policy, F. D. Maurice, and crucially separates him, even in his humanist phase, from the likes of Matthew Arnold – you cannot imagine him ever defining God as 'a stream of tendency by which all things seek to fulfil the law of their being' or 'the enduring power, not ourselves, which makes for righteousness'.[17] Yet he is also insisting that biblical language is not to be understood in a literal way. His objection is to the translation of that language into different terms and, in this, Ruskin's Christianity is far more radical than that of the liberals and might be thought to look towards a modern kind of religion.

One of Ruskin's most distinguished admirers in the early twentieth century was W. R. Inge (1860–1954), Dean of St Paul's, controversialist and author of a great many books on the mystical and Neoplatonic traditions in Christianity. Inge is an extremely interesting writer, who seems – rather like Ruskin himself in the mid-twentieth century – to have drifted from the cultural centrality he deserves. He argued for a Christianity based on his 'growing conviction that spiritual things are spiritually discerned, and spiritually proved'. A faith of this kind, he wrote in 1926, 'need not be afraid of scientific progress', for the field of science is distinct from that of the spirit.[18] Though there can be no doubt that scientific progress terrified Ruskin, it is significant that Inge could include him in his modern version of Neoplatonism and see him as one of the great religious thinkers of modern times.[19] In the passage from *Modern Painters* IV that I have just quoted, Ruskin is suggesting that the accounts of Creation in Genesis or the Psalms or the book of Proverbs are written in the language of myth, and that the language of myth is distinct from the language of science. As far back as 1843 when he wrote his *Letters to a College Friend* to Edward Clayton (1:399–502), Ruskin had recognized that the Bible could no longer be regarded as literally the word of God, and in 1867, in his post-Evangelical phase, he examined the matter systematically in one of the letters of *Time and Tide* (17:347–51). He would not have had to go much further to be able to see a more or less Darwinian view of things as not intrinsically at odds with Christianity, for he had clearly come to recognize that biblical language – for much of the time, the language of myth – makes no attempt to describe the actual processes of nature. There were some for whom this did not seem a problem. It was the view that Tennyson arrived at – admittedly with difficulty – in his elegy *In Memoriam*. It was Charles Kingsley's view, arrived at with much less difficulty – indeed with a certain enthusiasm. And it was powerfully endorsed by the Rev. Stewart Headlam, founder of the Guild of St Matthew, who said in a sermon of 1879:

> Thank God that the scientific men have [...] shattered the idol of an infallible book, broken the fetters of a supposed divine code of rules; for so they have helped reveal Jesus Christ in his majesty. [...] He, we say, is the Word of God; he is inspiring you, encouraging you, strengthening you in your scientific studies; he is the *wisdom* in Lyell or in Darwin. [...] It gives us far grander notions of God to think of him making the world by his Spirit through the ages, than to think of him making it in a few days.[20]

Headlam's Guild of St Matthew is a movement in the Anglican Church which combines the Ritualist practices of Anglo-Catholicism with radical socialism. In its reading of the social gospel and in the high value it sets on aesthetic and, by implication, natural beauty, it was deeply affected by Ruskin. But it was not a movement that Ruskin could have endorsed. Deep familial prejudices against Anglo-Catholicism, which in the days of Pusey and Keble had been politically conservative, made Ruskin as resistant to its liturgical attractions as he was to liberal theology. It would have been tarred for him with two brushes: that of Newman on the one hand and that of Maurice on the other. This is a way of saying that Ruskin in his later years, 'scared back by the grim face of reality' and unable to recapture the certainties of his childhood, was none the less resistant to any offer of a path into the future.

And yet he made such offers to himself from time to time. Though in *Modern Painters* IV he may *sound* like a literalist, he is very far from being one. What he appears to be saying is that the further we move from the mythical language of the Bible, the more likely we are to lose the sense of a divine Creator. But that is not to accept that the earth was in fact created in the way the Bible describes it. That process must have been in some sense of the word – not necessarily Darwin's – evolutionary. What the biblical language does is remind us of the mystery – the starkly physical mystery – that science may describe but cannot explain. To know what happened at the big bang is not to know *why* it happened. The laws of nature, whether rationally explained by a Deist like the natural theologian William Paley or abstracted away by a liberal like Matthew Arnold, can never account for the beauty of a flower. As Gerard Manley Hopkins noted of a bluebell: 'I know the beauty of our Lord by it.' That is something Ruskin could easily have assented to.[21]

So, for Ruskin, the language of science and the language of myth appeared to be at odds. In Letter 5 of *Fors Clavigera*, written in 1871, Ruskin writes of a friend who has been attending some lectures on botany (27:82–5). From these lectures she has learnt, to her amazement, that there are 'only seven sorts of leaves', and then that the petals of a flower are really leaves as well. And finally: 'my friend told me that the lecturer said, "the object of his lectures would be entirely accomplished if he could convince his hearers that there was no such thing as a flower."' Ruskin responds with amused irony to each of these announcements, they being so contrary to his sense – as they would have been to Darwin's, I suspect – of the richness and variety of creation, but at the last statement he explodes:

> [I]n that sentence you have the most perfect and admirable summary given you of the general temper and purposes of modern science. It gives lectures on Botany, of which the object is to show that there is no such thing as a flower; on Humanity, to show that there is no such thing as a Man; and on Theology, to show that there is no such thing as a God. No such thing as a Man, but only a Mechanism; no such thing as a God, but only a series of forces. The two faiths are essentially one [...]

One sees quite clearly here how one thing leads to another. It is not the recognition of the plant as something that changes as it grows that Ruskin objects to, but the reduction of all that variety and beauty to a series of fewer and fewer categories and so, by a chain of cause and effect, of man to a blind and aimless mechanism. He goes on, indeed, to affirm the essential truth of this metamorphic view of nature:

> Some fifty years ago the poet Goethe discovered that all the parts of plants had a kind of common nature, and would change into one another. Now this was

> a true discovery, and a notable one [...] In a certain sense, therefore, you see the lecturer was right. There are no such things as Flowers – there are only – gladdened leaves.

This is a passage of such grandeur that it is impossible to do justice to it in so brief an essay as this. The reader will easily gather, though, the direction it is moving in. The lecturer was at one and the same time right – as we have just seen – and

> in the deepest sense of all [...] to the extremity of wrongness, wrong. For leaf, and root, and fruit exist, all of them, only – that there may be flowers. He disregarded the life and passion of the creature, which were its essence [...]
>
> Now in exactly the sense that modern Science declares that there is no such thing as a Flower, it has declared there is no such thing as a Man, but only a transitional form of Ascidians and apes. It may, or may not be true – it is not of the smallest consequence whether it be or not. The real fact is, that, rightly seen with human eyes, there is nothing else but man; that all animals and things beside him are only made that they may change into him; that the world truly exists only in the presence of Man, acts only in the passion of Man.[22]

I cannot pursue every one of the passage's implications. I merely note in passing that, in the course of it, Ruskin acknowledges not only that the different parts of plants have a common nature but that human beings *may* be descended from apes. The process does not matter, he says – wrongly, we may think, to the extremity of wrongness. What does matter, though, is that there should be a moral consciousness – a portion of divinity – to witness the beauty and variety of things.

Notes

1. *The Works of John Ruskin*, ed. E. T. Cook and Alexander Wedderburn, 39 vols (London: George Allen, 1903–12), 7:205. Future references to this edition will be bracketed in the text in the form '7:205'.
2. The quotation is not exact, which suggests that Ruskin is – as is often the case – quoting from memory.
3. The essays were published as a book in 1862.
4. Derrick Leon, *Ruskin, the Great Victorian* (London: Routledge, 1949), 47. For much of what I say here about Buckland, I am indebted to Van Akin Burd, 'Ruskin and his "Good Master", William Buckland', *Victorian Literature and Culture* 36 (2008): 299–315.
5. Psalm 95:5. See note 16 below.
6. In *The Stones of Venice*, vol. 2, *The Sea-Stories* (London: Smith, Elder, 1853): see *Works of Ruskin*, 10:180–269.

7 The whole passage on imperfection is profoundly relevant to the present study. See *Works of Ruskin*, 10:189–204.
8 I am indebted for this insight to my friend Howard Hull.
9 Charles Darwin, *A Monograph on the Sub-class Cirripedia, with Figures of All Species: The Balanidae (or Sessile Cirripedes); The Verrucidae, etc., etc., etc.* (London: The Ray Society, 1854), plate 1 'Balanus Tintinnabulum', opposite p. 640.
10 The paper (read on 4 January 1837) was by Charles Darwin, 'Observations of Proofs of Recent Elevation on the Coast of Chili' [*sic*]: see *Proceedings of the Geological Society* 2 (November 1833 – June 1838, 446): 36:9n.
11 The attack on Whistler is in Letter 79 of *Fors Clavigera: Letters to the Workmen and Labourers of Great Britain*, vol. 7 (1877), *Works of Ruskin*, 29:146–69; the attack on Dickens (whose novels Ruskin generally admired) in *Fiction, Fair and Foul* (1880), *Works of Ruskin*, 34:264–399.
12 'Ruskin, himself, is a very simple matter. In face, in manner, in talk, in mind, he is weakness pure and simple. I use the word, not invidiously, but scientifically. He has the beauties of his defects; but to see him only confirms the suspicion given by his writing, that he has been scared back by the grim face of reality into the world of unreason and illusion, and that he wanders there without compass or guide – or any light save the fitful flashes of his beautiful genius.' *Henry James Letters*, ed. Leon Edel, 4 vols (London: Macmillan, 1974–80), 1:103.
13 John Milton, *Paradise Lost*, ed. Stephen Orgel and Jonathan Goldberg (Oxford: Oxford University Press, 2008), 3.
14 *The Ruskin Family Letters: The Correspondence of John James Ruskin, His Wife, and Their Son, John, 1801–1843*, ed. Van Akin Burd, 2 vols (Ithaca and London: Cornell University Press, 1973), 2:584.
15 Psalm 18:9, 11. Ruskin conflates two verses, perhaps misremembering a passage he once learnt.
16 Psalm 95:5 (quoted in *Works of Ruskin*, 6:109). In the Church of England, this Psalm, known as the 'Venite', was sung every Sunday morning at the service of Matins. Ruskin quotes not from the Bible but from the version in *The Book of Common Prayer* used in that service.
17 Matthew Arnold, *Literature and Dogma: An Essay towards a Better Apprehension of the Bible* (1973); reprinted in *The Complete Prose Works of Matthew Arnold*, ed. R. H. Super, 11 vols (Ann Arbor: University of Michigan, 1960–77), 6:189, 200 (Arnold's italics).
18 W. R. Inge, *The Platonic Tradition in English Religious Thought* (London: Longmans, 1926), 113–15.
19 Ibid., 90–95.
20 Quoted in Alec R. Vidler, *The Church in an Age of Revolution: 1789 to the Present Day* (Harmondsworth: Penguin, 1961), 119.
21 The quotation is from a journal entry dated 14 May 1870. Here is the full context: 'I do not think I have ever seen anything more beautiful than the bluebell I have been looking at. I know the beauty of our Lord by it. It[s inscape] is [mixed of] strength and grace, like an ash [tree]. The head is strongly drawn over [backwards] and arched like a cutwater [drawing itself back from the line of a keel]. The lines of the bell strike and overlie this, rayed but not symmetrically, some lie parallel. They look steely against [the] paper, the shades lying between the bells and the cockled petal-ends and nursing up the precision of their distinctness, the petal-ends themselves being delicately lit. Then there is the straightness of the trumpets in the bells softened by the slight entasis and [by] the square splay of the mouth. One bell, the lowest, some way detached and

carried on a longer footstalk, touched out with the tips of the petals an oval/not like the rest in a plane perpendicular to the axis of the bell but a little atilt, and so with [the] square-in-rounding turns of the petals.' *The Journals and Papers of Gerard Manley Hopkins*, ed. Humphry House and Graham Storey (London: Oxford University Press, 1959), 199. (The square brackets are Hopkins's.) This passage illustrates the extent of Ruskin's influence on Hopkins. The combination of close 'scientific' observation, draughtsman-like in its precision, with contemplative, almost mystical, reflection, bears comparison with Ruskin, though Hopkins's prose style is more idiosyncratic.

22 *Works of Ruskin*, 27:82–5.

Chapter 8

DARWIN AND THE ART OF PARADOX*

George Levine

Earlier chapters have dealt with Tennyson's ideas and anguish over evolutionary debates in *The Princess*, with the struggle between progress and destitution in *Locksley Hall* and with biographical links between Tennyson and Darwin. I want to shift attention in this chapter away from poetry and towards the form and language of prose. I want to celebrate Darwin as a writer whose vision and whose way of handling language had a profound effect on a vast range of literature beyond the poetic, and I want to do so primarily by attending to the element of surprise and paradox in his prose and then looking to some of the less frequently considered aesthetic consequences of taking seriously Darwin's way of looking at the world. The form of Darwin's thought and language can be detected in some unlikely and ostensibly unscientific places, where subjectivity and aesthetic value displace the quest for 'truth' and the 'full look at the worst' that we usually associate with his enterprise.

To do this I want to attend to Darwin the writer, who is not the dark messenger of disenchantment – 'the Devil's chaplain', draining meaning from the world, maddening John Ruskin with his materialist preoccupation with sex, and threatening Tennyson with the abolition of God from the universe. Rather, though he depicts a world of relentless and mindless process, he fills it with variety, wonder and meaning, opens a new (and yes, primarily secular) sense of the richness of life, encourages new ways of imagining and inspires new forms. He drives sombre poets like Tennyson and sober prose writers like Hardy into a vision of a world we never made and a nature, that is indeed red in tooth and claw, but he also inspires writers to turn inward, to value the one

---

* This chapter was first given as a plenary paper at Anglia Ruskin University in Cambridge, on 17 October 2009, at a conference on 'Darwin, Tennyson and Their Readers'. Much of the material has since been published in George Levine's *Darwin the Writer* (Oxford: Oxford University Press, 2011). We thank Oxford University Press for giving permission for the original paper to be published as a chapter in the present collection.

consciousness in the natural world that can recognize both consciousness and its absence. These qualities manifest themselves not so much in the philosophy implicit in or derivable from the idea of descent by modification through natural selection but in Darwin's way of seeing and imagining and arguing.

Near the end of the *Origin*, anxious about the reception he might receive, Darwin confesses that he by no means expects 'to convince experienced naturalists whose minds are stocked with a multitude of facts all viewed, during a long course of years, from a point of view directly opposite to mine'. But, he says, 'I look with confidence to the future, to young and rising naturalists, who will be able to view both sides of the question with impartiality.'[1]

Darwin underestimated the rapidity with which his ideas would become part of both scientific and literary culture, but he was right to look toward the future; fully to grasp the art of Darwin's prose requires a very modernist shift in point of view. Darwinian things as they are can only be perceived (or 'created') by way of changing perspectives, so that common sense about how we think and about what lies beyond us in the natural world no longer remains unquestioned. Darwin's new sublime is not so much outside, in the wonders of nature he so much admired and felt, but inside, in the power of the mind to imagine beyond what it sees. The shift partly cuts the bottom out of a language that attempted to describe a 'natural system' and seemed to reflect nature as it was.

As Gillian Beer has classically shown,[2] the language Darwin used and resisted, which is also the language we must continue to use, always implies agency, and it is in addition intrinsically anthropocentric – two qualities that Darwin laboured hard to reject. Moreover, nouns imply firm boundaries and absolutely distinguishable entities. It takes a lot of language to overcome the implications of language, and to do so Darwin developed a prose that often took the form of paradox. That paradoxical form is at the heart of the aesthetic turn at the end of the nineteenth century and in modernist literature. 'Natural history', said G. H. Lewes in 1860, 'is full of paradoxes.'[3]

We might, as literary scholars, want to attend first to the drive of Darwin's prose not only to make the rock-solid scientific case but (against his awareness of how hard this would be to achieve) through its rhetoric, to change our sense of probability and ultimately to change sensibilities. 'I remember too well', he wrote to a sceptical reader, 'how many long years my conversion took,' and that word 'conversion' carries a great deal of important weight.[4] Darwin's decades-long struggle to overcome the limits of his own common sense and instincts suggests how fundamental a shift in worldview his theory entailed.

His power to shift perspective was enabled, first of all, by the only exceptional talent he unequivocally attributed to himself in his autobiography – the power of observation. But observation, for Darwin, was really indistinguishable from thinking. When he 'looks' (a word with which the book just about begins), he sees contexts both of time and of space, and this contextual vision allows immediate recognition of anomaly and strangeness. His mind, even as early as the *Beagle* voyage, was stuffed with knowledge, which was always for him part of the very act of perception. Although he asserts that he worked on true Baconian principles, he tells us more revealingly – and truthfully – about hypotheses, 'I cannot resist forming one on every subject.'[5] Or, as he wrote to Wallace in 1857, 'I am a firm believer, that without speculation there is no good & original observation.'[6] His theory developed through a long accumulation of facts, hypotheses, hypotheses rejected, ideas solidified. It was not, certainly, a single, sudden revelation.

And yet the *Origin* itself, however cautious, however much it doubles back on itself and exploits conventional ways of seeing and thinking, gives the *feeling* of sudden revelation. We all know and smile at Huxley's reported response to his reading of the *Origin*: 'how extremely stupid not to have thought of that'.[7] But consider what that exclamation tells us about the book's argument, and its form: the *Origin* turns the world upside down, yet it makes common sense of the new vision. Traditionally sensible interpretations of the world became, to converts like Huxley, stupidity, and the *new* common sense was radically counterintuitive; even now, as we plunge into the details of Darwin's language and metaphors, it remains counterintuitive.

Seeing has to be learned, and it extends beyond what has been and continues usually to be regarded as the simple passive registration of the visible into an almost instinctive entanglement with past sights, future possibilities, and present contexts, an invisible network of connections historical and contemporaneous, and into the form of argument itself. As one member of the future generation, W. K. Clifford, made clear, seeing leads us into paradox. In a famous essay on 'The Philosophy of the Pure Sciences', Clifford begins by describing what he and everyone else seems to see upon entering the auditorium and then dramatically reverses himself – 'And yet', he says, 'I think we shall find on a little reflection that none of these statements can by any possibility have been strictly true.'[8] Clifford reminds us of how we create what we see, of the way the structure of our own eyes shapes and determines what we 'see'. The auditorium that we are confidently looking at is a construction of our mind inferred from the limits of our visual powers. Unlike Darwin, who was rhetorically more subtle, Clifford showily performed paradox in his wonderful, dramatic, and, as William James called them, too 'robustious' essays and lectures. But those essays look back toward Darwin's counterintuitive vision while they point forward to the more demonstratively paradoxical modes of the *fin de siècle* and Oscar Wilde.

As one of the key figures in introducing non-Euclidean geometry to England, Clifford changed the world from the common sense of Euclid to one whose constant if imperceptible curvatures made parallel lines meet. He insisted that the most important condition of mental development was resistance to conventionalities; what was required was 'plasticity' of mind, and he concluded his essay with a determined paradox in the mode of late-century iconoclasm: 'It is not right to be proper.'[9] We do nothing all day, says Clifford, in another essay, but change our mind. Such attitudes, paradoxical, anti-conventional, counterintuitive, are the marks of Darwin's thought and of his prose; they might also be understood as a literary transition from Victorianism to modernism.

Of course this is not all Darwin's doing. But Darwin had already taught us to break through conventions of perception, to see connections not literally visible to the eye, to infer history from static fact and movement from apparent stability. While taxonomy seemed to depend upon the way organisms shared characteristics, nothing was more valuable to Darwin's general theory than the singular, the aberrant, the anomalous, the exceptional, the rudimentary, the vestigial. 'Individual difference, though of small interest to the systematist', says Darwin, is 'of high importance for us, as being the first step towards such slight varieties as are barely thought worth recording in works on natural history'.[10] Darwin's science entailed a reversal, finding that it is not what is useful and important to organisms that allows us to understand their taxonomic place, for the useful and important will have been shaped by natural selection; rather, our best indication of genealogy is what is *not* useful and thus untouched by natural selection and need. To get where Darwin wants to take us, we have to recognize that his science is precisely *not* common sense, not habit, but a trip through the looking-glass.

From this perspective, Darwin's entire theory and all of its details, however soberly registered, amount to a giant paradox. What is stable is in motion; what is enormous depends upon minutiae; what seems peaceful is at war; struggle is often mutual dependency; lowly worms create the large green expanses of England; 6,000 years is no time at all; the term 'species' is an arbitrary one, given for convenience, and not essentially different from the term 'variety'; if unchecked by natural selection, even slow-breeding elephants would entirely cover the earth within five centuries; there are woodpeckers living where not a tree grows; there are web-footed birds that never go near the water; we are related physiologically to all living things, not only apes but barnacles and spiders. 'We behold the face of nature bright with gladness [...] we forget that the birds which are idly singing round us mostly live on insects or seeds, and are thus constantly destroying life' (61). The world of brilliant adaptations is moved not by a creative intelligence but by 'unknown laws of nature', and in nature itself the only intelligence is that of organisms, most obviously and particularly humans. 'There seems to be no more *design* in the variability of organic beings and in the action of natural selection, than in the course which the wind blows.'[11] 'It is a truly wonderful fact – the wonder of which we are apt to overlook from familiarity – that all animals and all plants throughout all time and space should be related to each other in group subordinate to group.'[12] The breathtaking litany could go on. Darwin's world emerges strange, unpredictable, sometimes comically perverse, sometimes awesome and terrifying. Seeing it Darwin's way requires a Huxleyesque revelation. Alice, once through the looking-glass, notes that 'what could be seen from the old room was quite common and uninteresting, but that all the rest was as different as possible.'[13] How stupid not to have thought of that!

Poets have always, in their metaphorical visions, had something of this perverse, even comic capacity to see things from a different point of view – 'negative capability' is the label of choice. Tennyson's 'The Eagle' reveals such a vision. I am not arguing that it is directly influenced by Darwin, but that it exemplifies an aesthetic power reasonably associated with Darwin's writing. As Tennyson's eagle flies, 'The wrinkled sea beneath him crawls.'[14]

Because the poem is so conveniently short, it is ubiquitously anthologized, but for the same reason it is easy to lose sight of what a stunning vision – before the age of aeroplanes and moon-walks – this is. As Darwin was to place us inside the consciousness of the female argus pheasant, Tennyson gives us the eyes of an eagle observing, from the enormous heights at which it soars, what Matthew Arnold from another more human perspective called the 'unplumb'd salt, estranging sea'.[15] And where does the sublimity of this image lie? Not in the sea but in the eagle's perception of it. It is that perception, imagined and interpreted by the poet, that turns the traditional mythical site of birth and death into something like a bedsheet or an old shirt. Or perhaps, an aged human who 'crawls' and hugs the earth childishly.

Although at the moment Tennyson wrote the poem, Darwin was probably being made seasick by those wrinkles, the image provides a perfect, miniature representation of the kind of shift of perspective toward which Darwin laboured. It emerges from the literary and scientific culture that precipitated Darwin's work, and it implies one of Darwin's great artistic and scientific achievements – the imaginative power to think beyond the human. Moreover, it does so in a double movement characteristic of Darwin's writing, the radical juxtaposition of two incompatible conditions – the vast and the domestic. The sublime re-emerges from this juxtaposition by way of the extraordinary consciousness that is capable of holding them together.

Consider that much of the argument of the *Origin* develops from Darwin's commitment to shift perspective from the anthropocentricity of William Paley's natural theology, which, finding divine intention behind all phenomena, reads natural phenomena as adjusted to human needs, desires, and perceptions. But how does it feel to see the world from the eagle's point of view and come to understand that the eagle's eye view is as valid and valuable as the human's? Darwin must have learned something of this strategy from Charles Lyell, who in a wonderful passage of unexpected reversal early in the first volume of *Principles of Geology*, reminds readers of how constrained their understanding is by the limits of their merely human perspective.

> If we may be allowed so far to indulge the imagination, as to suppose a being, entirely confined to the nether world – 'some dusky, melancholy sprite' – like Umbriel, who could 'flit on sooty pinions to the central earth,' but who was never permitted to 'sully the fair face of light,' and emerge into the regions of water and of air; and if this being should busy himself in investigating the structure of the globe, he might frame theories the exact converse of those usually adopted by human philosophers.[16]

I have to be careful here not to allow Lyell's wonderfully cultivated prose to upstage Darwin's, but it is important to note that this preoccupation with the constraints of perspective is central to the scientific tradition that Darwin entered, and it is also worth noting that Tennyson knew and was much influenced by Lyell's reading of the earth. (It is, of course, Lyell, rather than Darwin, who lies behind 'Nature, red in tooth and claw'.) For all three writers, we can see, as Gillian Beer has taught us, Milton figured importantly. He is there too in the novelists. George Eliot evokes Uriel in attempting to find a perspective equal to the complex interdependencies of relationships in *Middlemarch*. Lyell evokes Umbriel to correct our perspectives. Tennyson sees with eagle eyes. And Darwin, in one of the brilliant sequences of the *Origin*, invokes those eyes to 'stagger' us (his word) into a recognition that sublime vision can grow from Lyellian gradualist causes (204). Think about how the world changes as our point of vantage changes. Think about the astonishing possibility that aquiline vision is simply a natural extension of the first light-sensitive tissue in some lower organism. Darwin, like Tennyson, thrusts us into the sublime along pathways of domesticity.

There is another aspect to this Darwinian conjoining of the domestic with awesome vastnesses, one that almost certainly Darwin was not interested in encouraging, but that I believe had a very powerful effect on many writers who learned from him. That is, the shift of attention from the reality of nature, no longer laden with meaning and design, to the consciousness of humans, who emerge as the only real intelligences and creators of meaning, and whose capacity to 'see' in the full sense I've been exploring half creates that world. The true sublime is not the sea, or the mountains whose histories Darwin wonderfully explains and which leave him awestruck, but the narrative by which Darwin explains them and the intelligence that is capable of seeing the world in a grain of sand and filling it with a meaning that, in its raw material thingness, it doesn't have.

In pursuing this aspect of Darwin's work, I am taking up a suggestion made many years ago by A. Dwight Culler, who was dissatisfied with studies that attempt 'to trace the way […] writers derive from Darwin their views of nature, man, God, and society' because, as he claims, 'such studies do not seem to me quite to get at the heart of the problem.' Culler argues that the 'heart of the Darwinian revolution' is Darwin's 'dramatic reversal of orthodox thinking'.[17] Those counterintuitive aspects of Darwin's vision to which I have been pointing resonate through his prose and then through the writing of late Victorians and early modernists in ways that give us a different route to the effects of Darwin's art than the usual path through moral and religious struggles. Culler argues that, in fact, the basic form of Darwin's argument is not 'tragic', as Stanley Edgar Hyman insisted, but 'comic' in its radical reversal of our sense of things. What if we went from the naturalists' tragic rendering of humanity's relation to nature, to Oscar Wilde's?

The world of Darwin's prose isn't tragic, it's exhilarating. Everything 'means'; everything evokes a history and entanglements of relationship. The smallest wrinkled, crawling things – worms, ants, barnacles, or even pebbles on the beach – signify and imply narratives, inspire awe, evoke admiration. Darwin's art brings the world startlingly alive and fills it with meaning. David Kohn, concluding his discussion of modern developments in Darwinism, notes that 'Darwinism is blamed for taking meaning from the world by making divine purpose optional. But Darwinism in much of its practice is a project to populate the world with meaning, by identifying it in as many aspects of life as possible.'[18]

Darwinian explanation is necessarily narrative: producing hypotheses about every fact, Darwin turns the *Origin* into a series of convincing 'just-so stories', like those embedded in the famous treelike diagram of chapter 4. They are thought experiments, not pretending to describe what *is* or *was*, but hypothetical. This is what might have been. This is the most probable of possible stories.

A good place to start in locating Darwin's power to fill the world with meaning (and beneath the texture of the prose to imply the creative power of human intelligence) is in a little passage from his *Journal of Researches*, from the voyage of the *Beagle*, in which he talks of how the natives of Bahía Blanca attend to the 'rastro', or track, left by a fleeing enemy tribe:

> One glance at the rastro tells these people a whole history. Supposing they examine the track of a thousand horses, they will soon guess the number of

mounted ones by seeing how many have cantered; by the depth of the other impressions, whether any horses were loaded with cargoes; by the regularity of the footsteps, how far tired; by the manner in which the food has been cooked, whether the pursued travelled in haste; by the general appearance, how long it has been since they passed.[19]

This is both rhythmically impressive and an excellent representation of the way Darwin learned to look at the world. The force of the passage depends on an emotion not stated but implicit in its rhetoric: astonishment at the natives' capacity to read, at 'one glance', important meaning into visual evidence. Implicit is the natives' uncannily developed perceptiveness and intelligence, their power to create a logical and, indeed, useful sequence of thought out of footprints. Such a way of seeing the slightest details turns the world into a set of traces, but not merely of antiquarian value, for those traces reveal things profoundly important to the way the natives act at the moment. As nature seems to yield its secrets, our appreciation of the natives' power of half-perceiving, half-creating intensifies. Throughout the voyage and his career, Darwin reads the 'rastro' of nature itself, turning it into an endless set of stories. It is both a scientific and a poetic way to look at the world.

In tracing the nature of Darwin's art, one should begin with this almost instinctive transformation of a static or a single phenomenon into a history, a move that carries with it, almost inevitably, as metaphors tend to do, heavy emotional freight. Here, also from the *Journal of Researches*, is a geologizing Darwin considering the 'rastro'. He describes the 'din of rushing water' in a river in the Cordilleras and how the noise from 'the thousands and thousands' of stones as they 'rattled one over another' was heard night and day 'along the whole course of the torrent' and 'spoke eloquently to the geologist' as, with 'one dull uniform sound', they all hurried 'in one direction'. There follows what I think of as a characteristic double movement of his prose, a movement that explains both lucidly and literally the natural process and makes the merely natural awesome and wonderful:

> It is not possible for the mind to comprehend, except by a slow process, any effect which is produced by a cause repeated so often, that the multiplier itself conveys an idea, not more definite than the savage implies when he points to the hairs of his head. As often as I have seen beds of mud, sand, and shingle, accumulated to the thickness of many thousand feet, I have felt inclined to exclaim that causes, such as the present rivers and the present beaches, could never have ground down and produced such masses. But on the other hand, when listening to the rattling noise of these torrents, and calling to mind that whole races of animals have passed away from the face of the earth, and that during this whole period,

night and day, these stones have gone rattling onwards in their course, I have thought to myself, can any mountains, any continent, withstand such waste.[20]

Or withstand such a long, Ruskinian sentence. The Darwinian gaze connects immediately the 'mud, sand, and shingle' of the beaches with the insistent 'rattling noise' of those mountain torrents. In just this way, by recognizing connections, Darwin transforms the quotidian, the habitual, the trivially normal into the almost incomprehensibly vast and uncontainable. As Darwin sees and hears the little stones, he cannot help also recognizing their history and inferring their future. He cannot resist the implications for the mud many thousand feet deep, nor for what its depth implies about the extent of time of which – as Darwin connects the two – this landscape gives evidence. The excitement and beauty of the prose is in consequence of the stunning powers of connection revealed by the mind behind the prose. Fact and wonder intersect. After such a passage, no peaceful gurgling and rattling stream can be the same.

The double movement of Darwin's prose is the fundamental means by which he produces the reversal of perspective that I have been discussing. Assuming, via Lyell, that all phenomena can be explained by the working of causes now in operation, Darwin regularly presents to his audience and himself a stunning fact – the height of the Cordillera Mountains (brilliantly inferred later on to have been yet much higher millions of years before), the eagle's eye, the storage efficiency of hive-making bees, or the slave-making instincts of certain ants – and immediately expresses his awe and his recognition of the difficulty of explaining these phenomena naturalistically, according to his theory. But then he proceeds to do just that, invoking the most ordinary causes, *real* causes, whose efficacy we all can observe today. What he sees immediately seems incomprehensible; his prose implies a force beyond nature, but works at the same time to convince us that this force is *in* nature, and even ordinary and repetitive.

Or consider the double movement here, in another passage from the *Journal of Researches* explaining something we didn't know needed explanation. He notes how a quartz formation on the top of a large mountain abutting a 'sea-like plain' gives evidence of the working of waters no longer visible on the largely arid cliffs.[21] 'From custom', he says, 'one expects to see in the neighbourhood of a lofty and bold mountain, a broken country strewed over with huge fragments. Here nature shows that the last movement before the bed of the sea is changed into dry land may sometimes be one of tranquility.' With his already trained geological eye, Darwin has inferred unquestioningly the mountain's history, and the inference is confirmed as he casually notes how he 'was curious to observe how far from the parent rock any pebbles could be found'. He satisfies his curiosity and the readers', for he finds, 45 miles away, on the shore, quartz pebbles 'that certainly must have come from this source'.[22] To get the full feeling of Darwin's writing, one must register how much is implied in such a quiet discovery, how many days' travel it took to get from the peaks to the beach, how absolutely different the two environments, how profoundly imaginative the connection between little pebbles one takes for granted on a beach and the sublime mountains thousands of feet high, days and miles away. And one must try to feel the extraordinary satisfaction that comes with this kind of connection and knowledge – even the smallest pebble is endowed with a history almost as sublime as that of the mountain range.

The force of the story is once again not only in the history of the natural object but in the observer's powers to read it: that is, the act of counterintuitive discovery itself, the imagination that connects large and small through space and time. Much of the excitement and pleasure of Darwin's writing comes from the experience of surprise – how stupid not to have thought of that! – even in the midst of prose that struggles relentlessly to be rational and objective. (Note how often Darwin uses passive constructions in places where we probably wouldn't.) The reader responds not only to the sublime panorama in time and space but to the activity of the mind: both are startling.

In the literary world blossoming around Darwin, there is one most obvious place to look for this peculiar kind of startled recognition issuing in fascinating story-telling and celebration of the inquiring mind: the very Victorian genre of the detective novel, echoed and technologized in our contemporary TV police shows that spend half their time in labs where the slightest clues – a hair, a pin, a shred of cloth, mud caught in shoes, a daub of paint – all become means to tell the story of the crime and identify the culprit. I keep asking myself, as I watch those scenes in police labs with test tubes and tweezers and high tech machines, why are people watching this? Why am *I* watching this? What is it that virtually guarantees success to any narrative that can show a detecting mind brilliantly winnowing meaning out of large collections of fact? Sergeant Bucket in *Bleak House* already evoked Dickens's admiration. Wilkie Collins was already well at work on this sort of tracing the 'rastro', but Sherlock Holmes was the figure who most captured the public imagination.

In a 1907 book, *Through the Magic Door*, designed to encourage young people to pursue intellectual work, Holmes's creator Arthur Conan Doyle takes Darwin as a model, noting how reading *The Voyage of the Beagle* revealed to him immediately the amazing quality of Darwin's mind: 'Any discerning eye must have detected long before the "Origin of Species" appeared, simply on the strength of this book of travel, that a brain of the first order, united with many rare qualities of character, had arisen. Never was there a more comprehensive mind.

Nothing was too small and nothing too great for its alert observation [...] How a youth of Darwin's age [...] could have [...] acquired such a mass of information fills one with the same wonder, and is perhaps, of the same nature, as the boy musician who exhibits by instinct the touch of the master.'[23]

The qualities Doyle admires in Darwin are embodied in Holmes, who has captured the public imagination because (putting aside his arrogance and his drugs) he was so Darwinian, so startling in his power to juxtapose and make a story of unlikely phenomena and the facts of ordinary life. We first meet Holmes in a story in which Watson discovers an essay that Holmes has written about 'how much an observant man might learn by an accurate and systematic examination of all that came in his way':

> From a drop of water [...] a logician could infer the possibility of an Atlantic or a Niagara without having seen or heard of one or the other. So all life is a great chain, the nature of which is known whenever we are shown a single link of it. Like all other arts, the Science of Deduction and Analysis is one which can only be acquired by long and patient study, nor is life long enough to allow any mortal to attain the highest possible perfection in it. Before turning to those moral and mental aspects of the matter which present the greatest difficulties, let the inquirer begin by mastering more elementary problems. Let him, on meeting a fellow-mortal, learn at a glance to distinguish the history of the man, and the trade or profession to which he belongs [...] By a man's finger-nails, by his coat-sleeve, by his boots, by his trouser-knees, by the callosities of his forefinger and thumb, by his expression, by his shirt-cuffs – by each of these things a man's calling is plainly revealed. That all united should fail to enlighten the competent inquirer in any case is almost inconceivable.[24]

'What ineffable twaddle,' responds Dr Watson.[25] Precisely the sort of response Darwin feared (and got from John Herschel's initial reported response to the idea of natural selection – 'the law of higgledy piggledy').[26] The article seems to Watson a tissue of paradoxes, but Holmes proceeds, reading the conditions of Watson's old watch into an accurate life narrative. So Holmes becomes a popular hero on Cliffordian lines, just because he shakes Watson and his audience out of conventional thinking, demonstrates that everything is connected (everything 'means') and manifests those extraordinary powers that Doyle attributed to Darwin.[27]

Playing with the counterintuitive – which inevitably takes the form of paradox – is of course playing with fire, the kind of ambiguous and startling fire that marks modernist literature's ironic reversal of Victorian values. Certainly the directions *fin-de-siècle* writers took in exploiting the counterintuitiveness of Darwin seem to be at odds with Darwin's sense of personal life and decorum.

Clifford, clearly not a member of any aesthetes' circle, was already there. From a counterintuitive science he attacks mere propriety and habit, as Pater was doing at the same time from his absorption in philosophy, literature and, indeed, science. Gowan Dawson has recently shown that Victorians themselves made the connection, that Darwin – whose work makes genealogy and thus also sexuality the primary object of his attention – had to defend himself in his respectability against the connection;[28] but the quickest look at Pater and Wilde, for example, makes clear that Darwin's way of seeing lay behind aestheticism as much as it did behind bleak naturalism and the detective novel.

Who more than Walter Pater, among the Victorians, is better known for a prose that emphasizes the ephemeral nature of all things and is dedicated to shifting perspective to the inside, to sensibility? We know that Pater thought much about the development theory, although he saw Darwin's emphasis on change as only the most influential modern manifestation of a tradition going back to Heraclitus and developed in Hegel. He notes in *Plato and Platonism* that for Darwin, '"type" itself properly *is* not but is only always *becoming*,' and that "the idea of development […] is at last invading one by one, as the secret of their explanation, all the products of mind, the very mind itself, the abstract reason; our certainty, for instance, that two and two makes four.'[29] And who if not Darwin lies behind the famous opening to Pater's essay on Coleridge:

> Nature, which by one law of development evolves ideas, hypotheses, modes of inward life, and represses them in turn, has in this way provided that the earlier growth should propel its fibres into the latter, and so transmit the whole of its forces in unbroken continuity.[30]

For Darwin, 'differences blend into each other in an insensible series; and a series impresses the mind with the idea of an actual passage'.[31] In his understanding of this Darwinian world, Pater makes the next step: 'Things pass into their opposite.'

So Pater's enlisting of Darwin in great modernist developments – challenges to essentialism, to conventions of stability, and to the very concept of selfhood – seems right. Pater's prose might be thought of as a set of experiments in developing the Darwinian challenge to devise a language that resists its own fundamental nature and the worldview it inevitably implies. His famous hostility to formulae and essences, his insistence on the relativity of knowledge and his recurrence to 'inexpressible refinements of change',[32] as opposed to Darwin's 'imperceptible gradations',[33] makes Pater's almost ethereal and subjectivized world Darwin's world as well. The solidity and permanence of nouns, the clarity of borders, are hostile to Pater's vision and enterprise, which makes the centre of his work those aspects of Darwin's experience that led him to write, for example, 'Daily it is forced home on the mind of the geologist that nothing, not even the wind that blows, is so unstable as the level of the crust of this earth' (*Origin*, 323).

The effect is startling, most startling in that Darwin, so determined to make his argument on the strongest of empirical grounds and committed to ideals of scientific objectivity, ends by helping produce, through all the cultures of late nineteenth-century art, a sense not only of ephemerality but of the difficulty of really knowing what we are looking at, since it is certainly not the solid noun-like thing our eyes and brains are making it. And there emerges increasingly a rarefied intensification of subjectivity, a questioning of the nature of selfhood, and a new exploration of value. These qualities are most obvious in Pater's famous 'Conclusion' to the Renaissance when, following the lead of science, he claims that 'the service of philosophy [...] towards the human spirit is to rouse, to startle it into sharp and eager observation.'[34]

'Observation' is, as we recall, the great Darwinian power; the effect of good observation is, or should be, startling. Observation puts us onto traces, the 'rastro'; observation becomes a kind of story-telling; observation is indeed prophecy. Thinking, for example, of the way Thomas Hardy is normally enlisted to remind us of the bleak, tragic implications of Darwin's writing, I recall that powerful image in *The Woodlanders* (one of the most beautiful books of the century) in which Hardy describes Grace in Giles's cottage in the woods listening at night to the branches, which are disfigured with wounds resulting from their 'mutual rubbings and blows', from their 'wrestling for existence'. 'It was', Hardy notes, 'the struggle between these neighbours that she heard at night.'[35] There is Hardy in full Darwinian panoply, intensely observing but bleak as ever, and yet if one reflects on that image, one realizes that its startling power is not so much in the grim idea of universal struggle as in its representation of the trees as feeling beings, as 'neighbours' who have harmed each other but who speak their pains and their struggles.

Hardy's pessimism is sometimes a bit too easy, but his extraordinary sensibility to life and pain is clearly post-Darwinian. Some of his most precisely observed and beautiful images, as in *The Woodlanders*, are images of loss, pain, decay, but their aesthetic and moral power lies in the tenderness and sensibility of the language and in surprise at the revelation of the variety and possibility of forms of life:

> They went noiselessly over mats of starry moss, rustled through interspersed tracts of leaves, skirted trunks with spreading roots, whose mossed rinds made them like hands wearing green gloves, elbowed old elms and ashes with great forks, in which stood pools of water that overflowed on rainy days, and ran down their stems in green cascades. On older trees still than these huge lobes of fungi grew like lungs.[36]

Hardy will go on to suggest that these images provide evidence of the 'Unfulfilled Intention' and thus will make allegory out of these intensely observed and metaphorized particulars.

Darwin's job is more prosaic, it is true, but note how, in his famous simile of the 'tree of life', Darwin registers both the details of the literal growing tree and the metaphoric implications of its forms, and he manages to be both scientific and beautiful at the same time. The analogy covers a whole page and ends his chapter on 'Natural Selection':

> The green and budding twigs may represent existing species; and those produced during each former year may represent the long succession of extinct species. At each period of growth all the growing twigs have tried to branch out on all sides, and to overtop and kill the surrounding twigs and branches, in the same manner as species and groups of species have tried to overmaster other species in the great battle for life. The limbs divided into great branches, and these into lesser and lesser branches, were themselves once, when the tree was small, budding twigs; and this connexion of the former and present buds, by ramifying branches may well represent the classification of all extinct and living species in groups subordinate to groups. Of the many twigs which flourished when the tree was a mere bush, only two or three, now grown into great branches, yet survive and bear all the other branches; so with the species which lived during long-past geological periods, very few now have living and modified descendants. From the first growth of the tree, many a limb and branch has decayed and dropped off; and these lost branches of various sizes may represent those whole orders, families, and genera which have now no living representatives, and which are known to us only from having been found in a fossil state. As we here and there see a thin straggling branch springing from a fork low down in a tree, and which by some chance has been favoured and is still alive on its summit, so we occasionally see an animal like the Ornithorhynchus or Lepidosiren, which in some small degree connects by its affinities two large branches of life, and which has apparently been saved from fatal competition by having inhabited a protected station. As buds give rise by growth to fresh buds, and these, if vigorous, branch out and overtop on all sides many a feebler branch, so by generation I believe it has been with the great Tree of Life, which fills with its dead and broken branches the crust of the earth, and covers the surface with its ever branching and beautiful ramifications. (*Origin*, 113–14)

Here the word 'ramifications', etymologically precise if not obviously beautiful, sustains the parallel between the literal branching of a literal tree and the counterintuitive branching of varieties and species. Here, too, Darwin seems even to have attended to things like alliteration and internal rhyme, in the 'crust' covering the 'surface', in the 'broken branches', issuing in 'beautiful ramifications'. Hardy takes the shape of trees to the Unfulfilled Intention, Darwin to the condition of species; both see with startling precision and turn image into meaning.

Darwin, like Pater after him, refuses the merely habitual response, determined not to fall into what Pater would call our 'failure'; that is, our tendency to form habits, 'for habit is relative to a stereotyped world, and meantime it is only the roughness of the eye that makes any two persons, things, situations, seem alike' (250). Yes, this may seem like a stretch; Darwin would probably have been appalled by where Pater takes this rejection of habit, and – a man who was nothing if not proper – he would have been made unhappy by Clifford's 'it is not right to be proper.' But the connection is there and obvious and important. There is a wonderful phrase in the *Beagle* narrative, in which through the double movement of his prose, Darwin is explaining naturalistically the astonishing history of the Cordilleras. 'We must not now reverse the wonder, and doubt whether all-powerful time can grind down mountains – even the giant Cordillera – into gravel and mud' (260). Understanding how those magnificent mountains have dwindled under the pressure of 'all-powerful time' does not make them, or the process they undergo, or the mud and the gravel themselves, less wonderful. Darwin surely is one of the chief instigators to a revolution of perception that writers in the last third of the nineteenth century were absorbing into their ways of thinking and storytelling. They must not reverse the wonder!

But the aspect of Darwin's world that most upset habitual thinking was the idea that Daniel C. Dennett now most gleefully asserts – that nature and all those wonderful processes are mindless. He notes Darwin's 'shocking substitution of Absolute Ignorance for Absolute Wisdom in the creation of the biosphere [...] The very idea that all the works of human genius can be understood *in the end* to be mechanistically generated products of a cascade of generate-and-test algorithms arouses deep revulsion in many otherwise quite insightful, open-minded people.'[37] It might even have done so in Darwin himself, who we know was deeply influenced by Paley's natural theology and who, as Richards has shown, almost never uses mechanical analogies. At least in the first formulations of natural selection, Darwin implied his fundamental assumptions about the possibility of progress and perfection (thus, by the way, confirming Tennyson's hopeful reading). He was part of the audience to be shocked by where his 'reason' would take him. But I invoke Dennett here because his deliberately provocative (and historically misleading) way of formulating Darwin's dangerous idea emphasizes the fundamental paradox at the heart of the vision that has been most widely diffused through Western culture, and it is that paradox that I want to emphasize in considering, finally, the reach of Darwin's art into the work of late-century writers. Most people, John Hedley Brooke argues, rejected Darwin's sense of natural selection as benevolent.[38]

The *Origin* is pointed directly against a human-centered conception of nature and against the notion that design entails a designer, and in his autobiography Darwin is explicit: 'The old argument of design in nature', he writes, 'fails, now that the law of natural selection has been discovered' (50). The double movement of Darwin's prose is the stylistic enactment of this banishment of mind from nature: Darwin begins with wonder – 'He must be a dull man who can examine the exquisite structure of a [honey]comb, so beautifully adapted to its end, without enthusiastic admiration. […] Grant whatever instincts you please, and it seems at first quite inconceivable how [bees] can make all the necessary angles and planes, or even perceive when they are correctly made.' But then begins the explanation showing that it is not miraculous at all: 'the difficulty is not as great as it at first appears' (221). What follows, as Darwin invokes his 'great principle of gradation', is a consideration of a whole range of extant species and their various devices for holding honey. The leap, it turns out, is *not* catastrophist, but uniformitarian, as we learn that there are indeed grades of perfection in this honey-holding capacity. The explanation is long and detailed in the fascinatingly precise and yet speculative way Darwin can be, invoking geometry (with the advice of 'Professor Miller of Cambridge') to try to understand how the precisely efficient comb of hive bees might have developed from intentionless insects. He uses, as almost always, a synchronic analysis to explain the diachronic process. Darwin, like Holmes, moves from the awesome mystery of the hive bees' superb constructive powers, which would seem to imply some miraculous intelligent intervention, to the much more unsurprising activities of other bees. (Context is indispensable.) He sees 'real causes' now in operation and thus does not need to invoke an intelligent designer; without actually knowing the history of the hive bees as a species, he implies a history. The extraordinary becomes the product of the ordinary. The experience of the extraordinary remains, the shock of a world that can be seen as beautiful by human consciousness when the world itself is merely mindless (or instinctive) process:

> Thus, as I believe, the most wonderful of all known instincts, that of the hive-bee, can be explained by natural selection having taken advantage of numerous, successive, slight modifications of simpler instincts […] The bees, of course, no more knowing that they swept their spheres at one particular distance from each other, than they know what are the several angles of the hexagonal prisms and of the basal rhombic plates. The motive power of the process of natural selection having been economy of wax; that individual swarm which wasted least

honey in the secretion of wax, having succeeded best, and having transmitted by inheritance its newly acquired economical instinct to new swarms, which in their turn will have had the best chance of succeeding in the struggle for existence. (235)

We must not now reverse the wonder.

We are on the edge of *fin-de-siècle* irony here, a comedy that juxtaposes the brilliance of the conception against the mere blindly natural world that is being conceived. The bee passage stretches on for nine detailed and densely argued pages; as it withdraws intelligence from nature, it demonstrates it in the scientist/detective. It does not take much then to infer from Darwin's account that nature produces extraordinarily beautiful and complex structure while remaining dumbly incompatible with consciousness.

Commenting on the belief of some naturalists 'that very many structures have been created for beauty in the eyes of man, or for mere variety', Darwin claims dramatically: 'This doctrine, if true, would be absolutely fatal to my theory' (199). Nature is not interested in 'man', except where natural selection goes to work on him. But many theologians – not to speak of Calvin long before – regarded reliance on nature for evidences of God as dangerous. John Henry Newman didn't need Darwin to warn him off, having argued that 'it is a great question whether atheism is not as philosophically consistent with the phenomena of the physical world, taken by themselves, as the doctrine of a creative and governing power.'[39] After Darwin, however, nature, far from being the seat of justice, was as John Stuart Mill was to put it in *Nature*:

> In sober truth, nearly all things which men are hanged or imprisoned for doing to one another, are nature's every day performances [...] All this nature does with the most supercilious disregard both of mercy and of justice [...] Either it is right that we should kill because nature kills; torture because nature tortures; ruin and devastate because nature does the like; or we ought not to consider at all what nature does, but what it is good to do.[40]

Mill's argument separates morality from nature; modernism separates art from nature.

The tendencies I have been talking about, so prominently developed in Darwin, led in many directions, to the howling despair of James Thomson's 'City of Dreadful Night', to naturalist fiction and Hardy's cosmic ironies, but also to a new and intensified aestheticism. 'Science grows and beauty dwindles' says Tennyson's narrator,[41] but as science grew, the aesthetes and aesthetically oriented writers at the end of the nineteenth century took science's lessons and found paradoxically satisfying ways to the beautiful.

The increasingly inward turn of so much late-century writing seems to me to be precisely a response to the blankness of the material world. Darwin himself knew he had to struggle to explain the presence of consciousness in an undesigned and indifferent world, and in *The Descent of Man* he speculates on the processes by which this happens. The process, by the way, is precisely the one we have seen in his explanation of the eagle's eye and the hive-making bee – Daniel Dennett gleefully notes Darwin's 'shocking substitution of Absolute Ignorance for Absolute Wisdom in the creation of the biosphere'.[42] Then the establishment of a synchronic context – look how many kinds and degrees of consciousness there are in living animals; then the principle of imperceptible gradations – 'if no organic being except man had possessed any mental power, or if his powers had been of a wholly different nature from those of the lower animals, then we should never have been able to convince ourself that our high faculties had been gradually developed' (86); and thus the final banalizing and, from my point of view, Darwinian sublime explanation of the marvellous in terms of the ordinary. It all happens without a teleology, without a designer of the kind that even Alfred Wallace invoked because the marvel of human consciousness was just too extreme to be explained by natural selection. Wallace, alas, from Darwin's point of view, reversed the wonder.

But the mindlessness of a nature that produces a being capable of recognizing the mindlessness drives the action inside and shifts the burden of value from God to man. The contrast between a mindless and indifferent world and human sensibility and intelligence can produce tragic forms, it is true; but it is equally likely to produce comic ones, emphasizing the creative power of human imagination and intelligence, the inward turn, so effectively anticipated and spurred by Pater's *Renaissance*. The inward turn manifests itself most obviously in the fiction of Henry James, in the strange distortions, comic in their tragic implications, of Conrad's novels, where, in *Under Western Eyes*, Razumov scrawls 'Evolution, not Revolution' and lives out with bitter irony the impossibility of the ideal to which he aspires. But perhaps the best locus for an articulation both theoretically and performatively of the aesthetic, inward and paradoxical turn of post-Dawinian literature is in Oscar Wilde's remarkable, comic,and theoretically impressive dialogue – 'The Decay of Lying'.

Darwin turned the world on its head and tried to stay respectably quiet about it. Wilde spent all his time forcing people to notice its paradoxical nature and to laugh with it. Instead of the soul-raking 'Nature, red in tooth and claw', he has his main speaker, Vivian, complain, 'Nature is so uncomfortable. Grass is hard and lumpy and damp, and full of dreadful black insects.'[43] The difference in tone does not belie the similarity in understanding of nature. His speakers are elegant, snobbish and spoiled, chatting on about art with no attention to the sorts of needs that natural selection tends to. They seem hardly scientific, hardly the struggling protagonists of Gissingesque middle-class ordinariness or Hardyesque combatants against ill fortune or tortured Conradian grotesques. But despite being intent on delightful paradox and ostensibly trivial ironies, Vivan builds his theory of art out of Darwinian materials, laughing brilliantly along the way.

Consider this: 'My own experience is', says Vivian, 'that the more we study Art, the less we care for nature. What Art really reveals to us is Nature's lack of design, her curious crudities, her extraordinary monotony, her absolutely unfinished condition.'[44] Hardy's Unfulfilled Intention looms over this drawing room scene, but Hardy's sensibility can't get in. Against that deadly Unfulfilled Intention, Wilde builds up the fortress of Art, for Art, Vivian claims, 'is our spirited protest, our gallant attempt to teach Nature her proper place' (291). Wilde keeps laughing, but the ideas are as serious as the dark fatality that looms over Hardy and much nineteenth-century naturalism. Anticipating Culler's argument, he turns potential tragedy into comedy: 'Nothing is more evident than that Nature hates Mind. Thinking is the most unhealthy thing in the world, and people die of it just as they die of any other disease. Fortunately, in England at any rate, thought is not catching' (291).

Vivian's contempt for 'realism', the tediousness of the everyday, is an aspect of his sense of the incompatibility of mind and art with a nature that has no mind at all. The operations of natural selection are all directed at what is useful to the organism. Ruskin, Jonathan Smith has reminded us, was deeply upset by Darwin's concentration on the sexuality of organisms; he regarded such a view as consonant with the ugly utilitarian directions of modern capitalism. Flowers, Ruskin insisted, are not for reproduction but for the pleasure of human observers.[45] Wilde takes the Ruskinian perspective, pushing it to places Ruskin wouldn't have wanted to go and, of course, dropping Ruskin's high moralism. He has Vivian insist, simply, that 'the only beautiful things [...] are the things that do not concern us. As long as a thing is useful or necessary to us, or affects us in any way, either for pain or for pleasure, or appeals strongly to our sympathies, or is a vital part of the environment in which we live, it is outside the proper sphere of art.' (299) Of course, this is Wildean paradox, but it is a perfectly reasonable response to the threat of the utilitarian that natural selection and a nature merely functional and without consciousness pose.

Wilde's aestheticism affirms the Darwinian world in the act of rebelling against it. With God out of nature and design intrinsic only to the human, Wilde affirms art (deliberately outrageous because he calls it 'lying') as the antidote to the indifferent world and in effect puts it in place of the displaced God. The world is, through Wilde, filled with meaning, but only through human inventiveness, and the form he gives the theory is the paradox, the inversion of normal ways of thinking: 'One touch of Nature may make the whole world kin, but two touches of Nature will destroy any work of Art' (301). And Wordsworth, he claims, 'found in stones the sermons he had already hidden there,' a phrase tellingly indicative of the sense that it is not nature but human consciousness that creates meaning, (301).

Wilde's largest paradoxical move, of course, is to turn upside down the idea that art imitates life. 'Life, poor, probable, uninteresting human life – tired of repeating itself for the benefit of Mr. Herbert Spencer, scientific historians, and the compilers of statistics in general, will follow meekly after [the liar/artist] and try to produce, in her own simple and untutored way, some of the marvels of which he talks' (305). And when Vivian speaks of the Greeks' relation to art, his ironies imply a whole ethic as well as an aesthetic. 'They knew that Life gains from Art not merely spirituality, depth, thought and feeling, soul-turmoil or soul-peace, but that she can form herself on the very lines and colours of art and can reproduce the dignity of Pheidias as well as the grace of Praxiteles' (308). Wilde is as empiricist as Darwin, but his empiricism implies that experience registers reliably not the external world but one's own subjectivity; while we think we are observing objective reality, we are in effect mistaking our own impressions for the external world. It is human consciousness that puts meaning and order in the world. Art takes the place of the transcendental intelligent designer, and artful consciousness becomes the property exclusively of the human.

I don't for an instant want to suggest that this is where Darwin takes his own method, but I do want to insist that this kind of writing and thinking is recognizably a dramatic enforcement of the paradoxes that are so central to all of Darwin's major arguments. The form of paradox Wilde employs is more than a game. It suggests one way in which our culture has learned to handle the primal ignorance and indifference that Dennett discusses; that is, we read into nature the meanings and values that we claim to have derived from it. Feuerbach was there before this. But the position follows from Darwin's rendering of nature: it is human consciousness that fills the world with value and meaning, and for Wilde, the instrument of value is consciousness itself. It may well be a development out of natural selection for utilitarian purposes, but it becomes human just when it transcends utility. It certainly implies, as much as Hardy did, the incompatibility of nature and consciousness; but it celebrates consciousness and its ideal development, in Art.

'Where,' Vivian asks, 'if not from the Impressionists, do we get those wonderful brown fogs that come creeping down our streets, blurring the gas-lamps and changing the houses into monstrous shadows […] Things are because we see them, and what we see, and how we see it, depends on the Arts that have influenced us.' (312) In effect, Wilde is talking here, in the language of art, about what I have claimed Darwin did in the language of science. 'To look at a thing', Vivian says, in a formulation that recalls both Ruskin's claims about seeing and Darwin's practice, 'is very different from seeing a thing. One does not see anything until one sees its beauty.'[46] It may be that for Darwin, vision does not depend on the perception of beauty, but it is a reasonable argument to insist that Darwin's prose implies the beauty of nature, no matter how technical and 'objective'-seeming it becomes. Darwin's dogged pursuit of the meaning of his subjects implies a powerful engagement with them. And so he asks about everything: How account for its extraordinary qualities? Why is it here and not elsewhere? How did it get here? Why is it formed as it is? How is it related to other similar things encountered elsewhere? What elements of its nature are fundamental and inherited? What elements are shaped by the pressures of natural selection? These are not of course meanings that would matter in Wilde's theory. The virtuosity of Darwin's investigations has not quite the flair of Wilde's. But for both of them, seeing is a creative act, and the world emerges for Darwin as wonderful, staggering, awesome, sublime and somehow or other, explicable.

Still, it is a long way from Darwin's exhaustive effort to provide, through reason, for a probabilistic explanation of the improbable phenomena with which the world is populated. Wilde thrives on the improbable and therefore paradoxical; Darwin's writing is inflected by romanticism and the Victorian conventions that writers like Tennyson exploited so richly. Darwin softened up the world, as it were, for aestheticism and for Wilde, and he left us an art as dramatic as and more breathtaking than Wilde's, though a lot less funny. He left us a script for both tragedy and comedy, and I believe that the comedy needs, these days, a lot more emphasis. And so I want to close with just a touch of Darwin's encounters with the marvellous and wonderful and the moving way in which he assimilated his paradoxical vision of the world we never made, a world that drove many others to resistance.

In his chapter on the 'Struggle for Existence', we have a brilliant instance of the double movement of his prose, from the sublime to the ordinary and again to the sublime. There Darwin talks about an estate in Staffordshire (and in thus doing quietly brings his transmuting world home to his readers) in which a change in the planted part of the heath that has been enclosed was stunningly great, affecting all the forms of life there, insects, birds, plants. In an enclosed area, 'near Farnham, in Surrey', he found Scotch firs in astonishing abundance, and none in areas close by. The unlikeliness of this absolute difference provokes characteristic Darwinian investigation, and 'on looking closely between the stems of the heath', he finds,

> a multitude of seedling and little trees, which had been perpetually browsed down by the cattle. In one square yard, at a point some hundred yards distant from one of the old clumps, I counted thirty-two little trees; and one of them, judging from the rings of growth, had during the twenty-six years tried to raise its head above the stems of the heath, and had failed (72)

because of the grazing of the cattle. A forest in a square yard! And there is Darwin, on his knees, one suspects, counting the rings on the 'trunk' of a thwarted tree. As each detail, from the largest to the smallest, is registered, Darwin's imaginative reasoning confronts the wonder, explains it away, and leaves us awed at the total connectedness of things he has demonstrated.

In effect, Darwin's becomes a Rube Goldberg world in which his visual and experiential imagination expresses itself in the storytelling of mind experiment. A paragraph later he springs another counterintuitive vision: 'insects', he claims, 'determine the existence of cattle.' The flies are kept under control by certain kinds of birds, which would affect the propagation of the cattle in whose navels the insects breed, which would alter the vegetation, which would in turn affect the insects again, 'in ever-increasing circles of complexity' (73). The connectedness of things is not Darwin's invention, of course, but no writer ever brought home with such power the implications of that idea, as he in effect gives birth to the ecological imagination, the recognition that the head-bone is connected to the neck bone, and one can't affect any one part without affecting all the others.

But nature, as Darwin tells us and as modern ecologists know, is never this simple, and his visionary method, developed so brilliantly on the *Beagle*, and which takes him at last to that famous metaphor of the 'entangled bank', brings us to another lovely earlier metaphor. At one point in the *Origin*, Darwin talks about the almost miraculous way in which, 'when an American forest is cut down, a very different vegetation springs up,' but somehow, 'the trees now growing on the ancient Indian mounds [...] display the same beautiful diversity and proportion of kinds as in the surrounding virgin forests':

> Throw up a handful of feathers, and all must fall to the ground according to definite laws; but how simple is this problem compared to the action and reaction of innumerable plants and animals which have determined, in the course of centuries, the proportional numbers and kinds of trees now growing on the old Indian ruins.[47]

Pay attention to the language – to analogy and metaphor, to the lovely flying feathers, to the energy and excitement of the movement between two apparently separate events, from the enclosure act to the disappearance of Scotch firs in certain areas, from the appearance of a quartz rock on a beach to the sublime energy of a mountain 45 miles distant. Early in *Bleak House*, Dickens asks, 'What connection can there be?'[48] What connection between an aristocratic house in the Wolds and the poverty, mud and fog of London streets? We need not go to the Victorian novel for revelations that make such connections. Darwin's prose fills the world with astonishing relationships and thus with endless possibilities of meaning; everything presses us with questions and has value; everything evokes feeling. Learning with Darwin how to see the world without expecting anything of it except conformity to laws of nature, artists and writers learned how to revalue the world anew and how to tell their paradoxical stories. The world in Darwin's prose tells us stories that we need to know and tells them with the feeling that emerges from awe, wonder, insatiable curiosity, the risking of paradox and a trust in the powers of the mind, by way of imaginative juxtaposition and respect for things as they are, to make the mindless and very beautiful world make sense. It turns out that Ruskin was right: seeing is a mode of art, and it is prophecy.

So is it mere eccentricity to be 'startled' by such prose into keener recognition of the excitement and energy of the life that surges from the visual through imagination into language, or to find value in and care for this world of puzzling lovely flying feathers? Is it disenchanting to discover, through imaginative reason, that the struggle with which Darwin is normally associated is just as often 'mutual aid', or to be dazzled by the order that emerges from such apparent randomness, or to feel the history of things in the minutest details of their appearance? Yes, many of the consequences of this complexity are not pleasant, but there is a comic story to be told from Darwin's way of seeing and arguing, and the world that Darwin bequeaths us, with its flying feathers and its ramifying branches and its rastro drawing us back literally billions of years, is a gift that artists and writers continue to exploit and explore.

Notes

1 *Charles Darwin's Beagle Diary*, ed. Richard Darwin Keynes (Cambridge: Cambridge University Press, 1988), 379.
2 In *Darwin's Plots* [1983], 3rd ed. (Cambridge: Cambridge University Press, 2009).
3 George Henry Lewes, *Studies in Animal Life* (London: Smith, Elder & Co., 1862), 162.
4 Letter to John Phillips, 11 November 1859, *The Correspondence of Charles Darwin*, 20 vols, ed. Frederick Burkhardt et al. (Cambridge: Cambridge University Press, 1991), 7:372.
5 Charles Darwin, *Autobiography*, ed. Nora Barlow (New York: W. W. Norton, 1969).

6 Letter to A. R. Wallace, 22 December 1857, *Correspondence of Darwin*, 6:514.
 7 Thomas Henry Huxley, *Life and Letters of Thomas Henry Huxley*, ed. Leonard Huxley, 2 vols (London: Macmillan, 1900), 1:183.
 8 W. K. Clifford, 'The Philosophy of the Pure Sciences', in *Lectures and Essays*, ed. Leslie Stephen and Sir Richard Pollock (London: Macmillan, 1901).
 9 W. K. Clifford, 'On Some of the Conditions of Mental Development', in *Lectures and Essays*, 80.
10 Charles Darwin, *On the Origin of Species* [1859], facsimile ed. (Cambridge, MA: Harvard University Press, 1964), 51.
11 Darwin, *Autobiography*, 87 (emphasis in the original).
12 Ibid., 112.
13 Lewis Carroll, *Alice in Wonderland*, ed. Donald J. Gray (New York: W. W. Norton, 1975), 112.
14 Alfred Tennyson, 'The Eagle', in *The Poems of Tennyson*, ed. Christopher Ricks (London: Longmans, 1969), 495.
15 Matthew Arnold, 'To Marguerite – continued', line 24.
16 Charles Lyell, *Principles of Geology* [1830–1833], 3 vols, facsimile ed. (Chicago: University of Chicago Press, 1990), 1:32.
17 A. Dwight Culler, 'The Darwinian Revolution and Literary Form', in *The Art of Victorian Prose*, ed. G. Levine and W. Madden (New York: Oxford University Press, 1968), 225–8.
18 David Kohn, *The Darwinian Heritage* (Princeton: Princeton University Press, 1985), 332.
19 Charles Darwin, *The Voyage of the Beagle* [1845] (New York: Doubleday and Co., 1962), 101–102.
20 Ibid., 318.
21 Ibid., 109.
22 Ibid.
23 Arthur Conan Doyle, *Through the Magic Door* [1907] (Pleasantville, NY: Akadine Press, 1999), 245–6.
24 Arthur Conan Doyle, *Sherlock Holmes: Complete Novels and Stories*, 2 vols (New York: Bantam Books, 1986), 1:14.
25 Ibid.
26 Attributed to John Herschel by Charles Darwin in a letter to Charles Lyell, 10 December 1859, *Darwin Correspondence Project* online, Letter 2575: http://www.darwinproject.ac.uk/entry-2575
27 Serendipitously, Doyle was born in 1859, the year that Darwin finally published *On the Origin of Species*.
28 Gowan Dawson, *Darwin, Literature and Victorian Respectability* (Cambridge: Cambridge University Press, 2007).
29 Walter Pater, *Plato and Platonism* (London: Macmillan, 1907), 19–21.
30 Walter Pater, 'Coleridge', in *Appreciations, With an Essay on Style* [1889] (London: Macmillan, 1922), 64.
31 Darwin, *Origin of Species*, 51.
32 Ibid., 66.
33 Charles Darwin, *The Descent of Man and Selection in Relation to Sex* (Princeton: Princeton University Press, 1981), 51.
34 Walter Pater, *The Renaissance: Studies in Art and Poetry* [1888] (London: Macmillan, 1973). This passage is the source of the word 'startle', which I have deliberately used at various points in this chapter.
35 Thomas Hardy, *The Woodlanders* (London: Penguin, 1998), 311.

36 Ibid., 52.
37 Daniel C. Dennett, *Darwin's Dangerous Idea: Evolution and the Meaning of Life* (New York: Simon and Schuster, 1995).
38 John Hedley Brooke, 'Darwin and Victorian Christianity', in *The Cambridge Companion to Darwin*, ed. Jonathan Hodge and Gregory Radick (Cambridge: Cambridge University Press, 2009), 198.
39 John Henry Newman, Oxford University Sermons – Sermon 10: 'Faith and Reason Contrasted as Habits of Mind', delivered at Epiphany 1839, para. 39. Online: http://www.newmanreader.org/works/oxford/sermon10.html (accessed 8 August 2013).
40 John Stuart Mill, *Nature, the Utility of Religion and Theism* (London: Rationalist Press, 1904), para. 21. Lancaster E-text prepared by the Philosophy Department at Lancaster University. Online: http://www.marxists.org/reference/archive/mill-john-stuart/1874/nature.htm (accessed 8 August 2013).
41 *Poems of Tennyson*, 1368, line 246.
42 Dennett, *Darwin's Dangerous Idea*, 48.
43 Oscar Wilde, 'The Decay of Lying', in *Critical Writings of Oscar Wilde*, ed. R. Ellmann (New York: Random House, 1968), 291.
44 Ibid., 312.
45 Darwin is very clear about the absurdity of this position, identifying exactly the position that Ruskin would take, although he is not talking about Ruskin when he discusses it. '[S]ome naturalists', he says, protest 'against the utilitarian doctrine that every detail of structure has been produced for the good of its possessor. They believe that very many structures have been created for beauty in the eyes of man, or for mere variety. This doctrine, if true, would be absolutely fatal to my theory.' (199)
46 Wilde, 'Decay of Lying', 307–8.
47 Charles Darwin, *On the Origin of Species*, ed. Gillian Beer (Harmondsworth: Penguin, 2008), 57–8.
48 Charles Dickens, *Bleak House*, chapter 16 (Harmondsworth: Penguin, 1985), 272.

Chapter 9

SYSTEMS AND EXTRAVAGANCE: DARWIN, MEREDITH, TENNYSON

Gillian Beer

The year 2009 was one of centenaries: Darwin and Tennyson, Handel, Mendelssohn, Haydn, Purcell: all were explored and feted. But there are others, less noticed now: Edward FitzGerald is one and, some way behind him, two writers – sometime friends – who died within weeks of each other in 1909, Algernon Swinburne (10 April) and George Meredith (18 May). Perhaps the rather faded reputations of those last three may suggest that we have lost touch with the extravagance and extremes that mattered so much in the latter part of the nineteenth century and that opened the way to modernism. In this chapter I shall explore a few of the paradoxical relations between systems and extravagance in later Victorian thinking and fiction. Just as the sublime is key to Romantic sensibility, extravagance is its transformed equivalent in subsequent generations.

In our culture, music has become astonishingly more available than it ever was in history before. All four centennial composers are heard daily in homes as well as in the concert hall. They are heard not in adaptations for the piano (as so often in the Victorian period) but in full orchestra, pouring out of speakers, or straight into our ears, downloaded. Music is scattered abroad, as snack music in lifts and restaurants, as extraordinarily exact re-imaginings of how the works would first have sounded with their original instruments. Literature, on the other hand, though opened up on the Internet, does not have quite that capacity to perform in public. There remains that still solitary act of reading, even in the midst of book clubs and radio, audio and author readings that have spread the means of receiving the word. The paradox about the act of reading is that it opens the single silent reader to the tumult of other lives, within and without, and to other times and places, though permeated by the world in which we read. There is an extravagance about the outcome, a frugality about the process.

But I don't want to draw the boundaries of my capacious title too wide: the pairing of systems and extravagance could be like the fictional *Revelations of Chaos* in Disraeli's *Tancred*, of which it's said, 'It explains everything, and is written in a very agreeable style.'[1] I have something more specific in mind. In particular, I want to get away from the implied dualism of the pairing, which may mislead us.

I suppose the example that will spring to most people's minds of the opposition between systems and extravagance in Victorian fiction will be Dickens's *Hard Times* (1854). There, disassembled 'facts' are ground up into a system that makes no room for imagination or fancy. The circus and its people, unruly, slipshod, but alive to the present and its emotions, are set over against the Gradgrind regime with its selfish insistence on material success. 'Utilitarianism', analytical, logical, and statistical in its methods, is represented in Dickens's novel as the system that authenticates this barren behaviour. There are paradoxes at work here. John Stuart Mill is the writer most associated with a reasoned utilitarianism, through his 1861 essay of that name. But in 1859 he had published his yet more famous essay 'On Liberty' and much later in his autobiography (1873) he reveals how after a childhood under the sway of his father's intellectual forcing 'system' he had a breakdown that began at the age of 20 (1826). The boy who was a skilled linguist, logician, and economist by his mid-teens discovered the desperately needed power of emotion through poetry, particularly the poetry of Wordsworth, which he read in the autumn of 1828. Music, especially melody, had earlier excited his enthusiasm, but now he entered and learnt to value 'states of feeling, and of thought coloured by feeling, under the excitement of beauty'. This was 'a source of inward joy, of sympathetic and imaginative pleasure, which could be shared by all human beings'.[2] Excitement, beauty, joy, pleasure, sharing: these are the emotions that Mill is now at liberty to explore in their extremes. In contrast to his friend Roebuck, he feels no animosity between that imaginative reach and an understanding of the physical laws of scientific fact:

> He [Roebuck] saw little good in any cultivation of the feelings, and none at all in cultivating them through the imagination, which he thought was only cultivating illusions. It was in vain I urged on him that the imaginative emotion which an idea, when vividly conceived, excites in us, is not an illusion but a fact, as real as any of the other qualities of objects; and far from implying anything erroneous and delusive in our mental apprehension of the object, is quite consistent with the most accurate knowledge and most perfect practical recognition of all its physical and intellectual laws and relations. The intensest feeling of the beauty of a cloud lighted by the setting sun, is no hindrance to my knowing that the cloud is vapour of water, subject to all the laws of vapours in a state of suspension; and

I am just as likely to allow for, and act on, these physical laws whenever there is occasion to do so, as if I had been incapable of perceiving any distinction between beauty and ugliness.[3]

Mill confounds the opposition between explanation and ecstasy and finds a balance beyond the conception of the people in *Hard Times*, and perhaps beyond Dickens as a writer. Wordsworth proved the key to his escape. Mill escaped from system and exceeded its bounds, but the temper of his experience is not that of exaggeration or extravagance. Indeed, he denies any such interpretation.

The Excursion did not speak so strongly to Mill as did 'Intimations of Immortality', yet at the core of the *Excursion*'s 350-odd pages there occurs a passage that speaks to Mill's insight. Wordsworth here pursues the meaning of 'excursion' out beyond expedition or wandering, into an activity of mind and heart that uncovers fresh possibilities, sustained by science enkindled:

> [...] Science then
> Shall be a precious visitant; and then,
> And only then, be worthy of her name:
> For then her heart shall kindle; her dull eye,
> Dull and inanimate, no more shall hang
> Chained to its object in brute slavery;
> But taught with patient interest to watch
> The processes of things, and serve the cause
> Of order and distinctness, not for this
> Shall it forget that its most noble use,
> Its most illustrious province, must be found
> In furnishing clear guidance, a support
> Not treacherous, to the mind's *excursive* power.[4]

Wordsworth italicizes that word 'excursive' – a curious word that negatively signifies 'aimless' but in Johnson's *Dictionary* also means 'beyond fixed limits'.[5] Again, the escape from fixed system is emphasized and reaching out is praised. But the examples I have so far called in do not quite embrace extravagance.

Darwin does. His theory on the origin of species by means of natural selection relies on a hyperbolic expansion of time backwards through aeons in which slow events have accumulated change as an outcome of variability. It relies also upon fecundity and profusion, 'the astonishingly rapid increase of various animals in a state of nature, when circumstances have been favourable during two or three following seasons'.[6] 'Look at the most vigorous species; by as much as it swarms in numbers, by so much will its tendency to increase be further increased' (67). That profusion is equalled (sometimes outdone) by the destruction experienced within each generation of organisms:

> Seedlings, also, are destroyed in vast numbers by various enemies; for instance on a ground three feet long and two wide, dug and cleared, and where there could be no choking from other plants, I marked all the seedlings of our native weeds as they came up, and out of the 357 no less than 295 were destroyed, chiefly by slugs and insects. (67)

There is no simple distinction, he shows, between organism and environment, since the environment is itself the interplay of multiple and often conflicting needs and desires among the many organisms that inhabit the common space.

> Let it be borne in mind how infinitely complex and close-fitting are the mutual relations of all organic beings to each other and to their physical conditions of life. (80)

So these relations are multiple, yet close-honed. They rely upon excess numbers of individuals for the species to survive. Waste is not wasteful. Without such profusion, extinction rapidly follows. And extinction, equally, is to be newly understood in the wake of Darwin as on a vast scale, embroiling almost all species that have earlier peopled the earth.

> Judging from the past, we may safely infer that not one living species will transmit its unaltered likeness to a distant futurity. And of the species now living very few will transmit progeny of any kind to a far distant futurity; for the manner in which all organic beings are grouped, shows that the greater number of species of each genus, and all the species of many genera, have left no descendants, but have become utterly extinct. (489)

Darwin was much affected by Malthus's insistence on the ill-matched growth of population beyond food supply. For Malthus proliferation was a threat to be combated, but his essay 'On Population' relies for the force of its argument on the overwhelming energy of procreation. Strikingly, his chosen example at the start of the essay comes from plant life – fennel, which would rapidly overwhelm the world were it not kept in check by other plants. Darwin's thinking with Malthus seizes on the rampaging energy of this example, which immediately takes the potential of the argument out beyond the human into the natural world at large. Malthus concentrated on the human dilemma; Darwin understood that humanity could only ever provide a minor example of the massive and massively complex processes at work among entangled organisms through the past, into the present and beyond.

The world imagined by Darwin demands stupendous acts of exaggeration, something way beyond the balanced conciliations of Mill, for example. Time turns into aeons, progeny spill unlikeness from the parent type. Sex is a great aid in diversification with its bringing together of two *unlike* genetic streams, instead of the replications of amoebas splitting or of parthenogenesis – virgin birth – or of hermaphroditic productions, though all these methods over time do also deliver change. Georges Cuvier had 60 years earlier introduced the concept of species extinction. Darwin expanded it: the struggle to survive is unrelenting, undergone by individuals and species alike. And almost all these profuse and interlocking life forms also vanish over time into extinction. This is a story that makes Wagner's *Ring Cycle* look very small scale.[7]

The famous concluding sentences of the *Origin* insist on improbability even as they assert the extraordinarily contradictory outcome:

> Thus, from the war of nature, from famine and death, the most exalted object which we are capable of conceiving, namely, the production of the higher animals directly flows. There is grandeur in this view of life, with its several powers, having been originally breathed into a few forms or into one; and that, whilst this planet has gone cycling on according to the fixed law of gravity, from so simple a beginning endless forms most beautiful and most wonderful have been, and are being, evolved. (490)

Grandeur and exaltation, beauty and wonder, and at base – a few forms and simplicity. Grandeur emerging too from dire conditions: famine, warfare, death. The shifts in scale here match Darwin's methods – working assiduously, with scrupulous observation of lowly plants and creatures, with minute differences, with almost imperceptible change, yet exploding into an argument that reaches back into an inconceivable past and that calls in evidence drawn from varying phases of that past alongside the plenitude of the natural world now.

The word 'endless' is a key ('endless forms most beautiful and most wonderful'). Darwin draws it from biblical sources and from the work of natural theologians but turns it to his own needs. Indeed, his work seems directly to challenge some biblical uses. In 1 Timothy 54:1–4, for example, St Paul comments ironically: 'Neither give heed to fables and endless genealogies, which minister questions, rather than godly edifying which is in faith.' This is work without the bounds of time. It must be true in all fields and in all circumstances if it is to have validity. It must work on a stupendous scale while acknowledging its humdrum origins. Later, at the turn of the twentieth century, Gustav Mahler in *Das Lied von der Erde* ('The Song of the Earth')[8] will sound a note like this: '*Ewig, ewig*' (endless, eternal, everlasting), though there 'for ever' is drenched in sadness instead of endowed with that recalcitrant upward lift of hope that Darwin finds even in the midst of famine and death.

Darwin's is a story that exhibits a system at work. But it is a system that demands extravagance. In his later years Darwin expanded on his theory of sexual selection as an aspect of natural selection. He had said that looking at a peacock's tail made him feel sick, so excessive, so out-of-scale with use did its elaboration appear to be. He sometimes felt puzzlement, even occasional exasperation, at the sheer inventiveness of forms in nature; of orchids Darwin writes:

> Hardly any fact has struck me so much as the endless diversities of structure, – the prodigality of resources, – for gaining the very same end, namely, the fertilisation of one flower by the pollen from another plant.[9]

His later years were spent seeking a system *implicit* in the inordinate, the decorative, the ornamental, in the drive of sexual desire, though more difficult to trace in the *self*-satisfaction of hermaphroditic slugs. Sexual selection demanded flaunting, display, theatre, extravagance, scents and song. The males of most species, his researches showed, were driven to display; the females were the choosers (though choice might sometimes be a false word to describe the process of accepting the successful male's advances). Beauty re-emerged as a key element in his enquiry, and he argued that humans were not the sole possessors of aesthetics and of delight in art. Birdsong was prior to language; it expressed territorial and erotic claims by means of all the pleasures of skilled elaboration.

Combining system and extravagance in Darwin's work releases fresh possibilities for writers (and for all human beings). Curiously, though, the two most famous phrases associated with his work are not from Darwin at all: one is 'the survival of the fittest', from Herbert Spencer[10]; the other, which notably precedes the *Origin*, is, of course, Tennyson's famous half line, 'Nature, red in tooth and claw', from *In Memoriam*.[11]

For Tennyson as for Darwin the writing of Charles Lyell was foundational. His geology demonstrated a world changing over time, encompassing earthquake and disaster, but moving in the main with extreme slow pace – and implacable. To Tennyson the extinction of species seemed more heartbreaking than it did to Darwin. In one late poem, 'The Islet', Tennyson seems also to have grasped the meagreness of isolation in Darwinian terms:[12]

'Whither O whither love shall we go,
For a score of sweet little summers or so?'
The sweet little wife of the singer said,
On the day that follow'd the day she was wed,
'Whither O whither love shall we go?'
And the singer shaking his curly head
Turn'd as he sat, and struck the keys
There at his right with a sudden crash,
Singing, 'and shall it be over the seas
With a crew that is neither rude nor rash,
But a bevy of Eroses apple-cheek'd,
In a shallop of crystal ivory-beak'd,
With a satin sail of a ruby glow,
To a sweet little Eden on earth that I know,
A mountain islet pointed and peak'd;
Waves on a diamond shingle dash,
Cataract brooks to the ocean run,
Fairily-delicate palaces shine
Mixt with myrtle and clad with vine,
And overstream'd and silvery-streak'd
With many a rivulet high against the Sun
The facets of the glorious mountain flash
Above the valleys of palm and pine.'

'Thither O thither, love, let us go.'

'No, no, no!
For in all that exquisite isle, my dear,
There is but one bird with a musical throat,
And his compass is but of a single note,
That it makes one weary to hear.'

'Mock me not! mock me not! love, let us go.'

'No, love, no.
For the bud ever breaks into bloom on the tree,
And a storm never wakes on the lonely sea,
And a worm is there in the lonely wood,
That pierces the liver and blackens the blood,
And makes it a sorrow to be.'[13]

Darwin's language persistently links the words 'new and improved' where for Tennyson, from *In Memoriam* through to 'Locksley Hall Sixty Years After', what is lost is not only irretrievable but perhaps better than what remains.

In Memoriam
LVI

'So careful of the type?' but no.
    From scarped cliff and quarried stone
    She cries, 'A thousand types are gone:
I care for nothing, all shall go.

'Thou makest thine appeal to me:
    I bring to life, I bring to death:
    The spirit does but mean the breath:
I know no more.' And he, shall he,

Man, her last work, who seem'd so fair,
    Such splendid purpose in his eyes,
    Who roll'd the psalm to wintry skies,
Who built him fanes of fruitless prayer,

Who trusted God was love indeed
    And love Creation's final law –
    Tho' Nature, red in tooth and claw
With ravine, shriek'd against his creed –

Who loved, who suffer'd countless ills,
    Who battled for the True, the Just,
    Be blown about the desert dust,
Or seal'd within the iron hills?

No more? A monster then, a dream,
    A discord. Dragons of the prime,
    That tare each other in their slime,
Were mellow music match'd with him.

O life as futile, then, as frail!
    O for thy voice to soothe and bless!
    What hope of answer, or redress?
Behind the veil, behind the veil.[14]

Most of Tennyson's publications came too late for the period of Darwin's heady delight in poetry. For Darwin the poets are Byron, Wordsworth, Thomson, Shelley, Keats. Once he was married and deep in his researches, his enthusiasm for poetry waned and then vanished. Yet it seems improbable that he should not have been aware of *In Memoriam* after its publication in 1850, so powerful and central was it in Victorian cultural life, and so painfully close was its reiterated searching of the wound of loss. Charles and Emma Darwin's 10-year-old daughter Annie died in April 1851. Darwin's theory relies, I have suggested, not only on extravagance of scale but on extravagance of loss:

> We behold the face of nature bright with gladness, we often see superabundance of food; we do not see, or we forget, that the birds which are idly singing round us mostly live on insects or seeds, and are thus constantly destroying life. (62)

For Tennyson the key forms that experience takes in language are repetition, iteration and ornament. The power of *In Memoriam*, like that of Schubert's *Winterreise* (1826), lies in its dogged refusal to let go of grief – the *impossibility* of letting it go because with its loss the loss of the beloved is complete. Within that compass there are persistent subtle iterations and shifts of mood in *In Memoriam*. The enclosure of the rhyme scheme *abba* – an endless return and turning back on itself, an ever expanding *doxa* (the glory of common belief) – prolong our encounter until it is no longer encounter but immersion. This is emotion fully known, and known at a scale beyond that of the individual alone. The rhyme retards and fulfils. This is very different from the two 'Locksley Hall' poems, particularly the urgent stamping of the metre and rhyming couplets in 'Locksley Hall Sixty Years After'. Here the impatience of the old charges the lines:

> Ay, for doubtless I am old, and think gray thoughts, for I am gray;
> After all the stormy changes shall we find a changeless May?[15]

In this late poem (1888) Tennyson does indeed scan science: evolution, medicine, commerce, seem at first to point towards a more perfect world, *once systems fail*:

> When the schemes and all the systems, Kingdoms and Republics fall,
> Something kindlier, higher, holier – all for each and each for all?
>
> All the full-brain, half-brain races, led by Justice, Love, and Truth;
> All the millions one at length with all the visions of my youth?

All diseases quench'd by Science, no man halt, or deaf or blind;
Stronger ever born of weaker, lustier body, larger mind?

Earth at last a warless world, a single race, a single tongue –
I have seen her far away – for is not Earth as yet so young? –

Every tiger madness muzzled, every serpent passion kill'd,
Every grim ravine a garden, every blazing desert till'd,

Robed in universal harvest up to either pole she smiles,
Universal ocean softly washing all her warless Isles.

Warless? when her tens are thousands, and her thousands millions, then –
All her harvest all too narrow – who can fancy warless men?

Warless? war will die out late then. Will it ever? late or soon?
Can it, till this outworn earth be dead as yon dead world the moon?[16]

Darwin and Malthus conjoin in this turbulent exasperated vision of the future, with its looming crisis of population, a future first idealized, exaggerated, then blotted out.

Tennyson gave early encouragement to George Meredith. In 1851, the year after the publication of *In Memoriam* and at a time when Tennyson was newly the Poet Laureate, he wrote to Meredith on the publication of Meredith's first long poem, the idyll 'Love in a Valley', saying that he wished he had written it himself and that 'he went about the house repeating its cadences to himself.'[17] Like Tennyson, Meredith was always a poet of sound and metre first.

Meredith's first major novel, *The Ordeal of Richard Feverel*, was published, like the *Origin*, in 1859. This is a novel that attacks the dictatorship of an educational system imposed by a father on his son. Sir Austin Feverel, embittered by his wife's adultery, is determined to bring his son Richard to manhood in a state of absolute sexual ignorance (or innocence) as well as intellectual and ethical excellence. Sir Austin's dour collection of aphorisms, 'The Pilgrim's Scrip', has been published anonymously, the first and crucial one being, 'I expect that woman will be the last thing civilized by Man.'[18] Sir Austin is described as a monomaniac, organizing the world into a series of dramas with Sir Austin 'pointing out to his friends the beneficial action of the System from beginning to end' (103). There is a tone of rancorous comedy in the ironies explored throughout the novel, and the rancour deepens in the latter part of the book where Sir Austin's interference blights the innocent love between Lucy and Richard, leading to the novel's tragic ending. The love between the 'natural' heroine Lucy and young Richard is presented with all the fervour and lyricism of which Meredith's writing is capable. At their first encounter the reader sees the girl before Richard does so, but she emerges as an aspect of a natural world rich in sounds, scents, sight, everything in motion:

> Above green-flashing plunges of a weir, and shaken by the thunder below, lilies, golden and white, were swaying at anchor among the reeds. Meadow-sweet hung from the banks thick with weed and trailing bramble, and there also hung a daughter of earth. (127)

As always in Meredith's early writing, the style ricochets between this lyricism and a dissonant comedy that refuses to stabilize, as if he fears that any resolution will kowtow to authority or tell less than he knows.

Something of the same instability and excess marks the actual *form* of his great 60-sonnet sequence 'Modern Love' a very few years later (1863), where the 14 lines of the sonnet are extended to 16 as if there is always too much, and too many different kinds, of feeling to be crammed into 14 lines. That forensic masterpiece takes us inside all the moods – chafed, triumphant, desolate, conspiratorial, aroused, depleted – of a marriage breaking up.

> By this he knew she wept with waking eyes:
> That, at his hand's light quiver by her head,
> The strange low sobs that shook their common bed
> Were called into her with a sharp surprise,
> And strangled mute, like little gaping snakes,
> Dreadfully venomous to him. She lay
> Stone-still, and the long darkness flowed away
> With muffled pulses. Then, as midnight makes
> Her giant heart of Memory and Tears
> Drink the pale drug of silence, and so beat
> Sleep's heavy measure, they from head to feet
> Were moveless, looking through their dead black years,
> By vain regret scrawled over the blank wall.
> Like sculptured effigies they might be seen
> Upon their marriage-tomb, the sword between;
> Each wishing for the sword that severs all.[19]

That first poem's impasse is followed by others that track very different moods: rancorous play and complicity among them. They give a dinner party and act the part of happily married pair:

> But here's the greater wonder; in that we
> Enamoured of an acting nought can tire,

> Each other, like true hypocrites, admire;
> Warm-lighted looks, Love's ephemerioe,
> Shoot gaily o'er the dishes and the wine.
> We waken envy of our happy lot.
> Fast, sweet, and golden, shows the marriage-knot.
> Dear guests, you now have seen Love's corpse-light shine.
> (140–41)

The final poem completes the yearning, the chafing sense of loss, diminution and entrapment. Lives here have succumbed to systems, and these systems are not simply external in the marriage vows but internalized.

> Thus piteously Love closed what he begat:
> The union of this ever-diverse pair!
> These two were rapid falcons in a snare,
> Condemned to do the flitting of the bat.
> Lovers beneath the singing sky of May,
> They wandered once; clear as the dew on flowers:
> But they fed not on the advancing hours:
> Their hearts held cravings for the buried day.
> Then each applied to each that fatal knife,
> Deep questioning, which probes to endless dole.
> Ah, what a dusty answer gets the soul
> When hot for certainties in this our life! –
> In tragic hints here see what evermore
> Moves dark as yonder midnight ocean's force,
> Thundering like ramping hosts of warrior horse,
> To throw that faint thin line upon the shore!
> (155)

Meredith devours his own life with a ferocious percipience in *Modern Love* and also to quite a large extent in *Richard Feverel*. He had found himself a single parent of a small boy after his wife eloped with Henry Wallis, the painter. He had been wrung by the pains of a marriage of passion and equality collapsing under the weight of poverty, miscarriages, and ill-matched ambitions. He had observed his own narcissism and tried to escape it. This sense of a life flung into the pot of creativity, flinching, acrimonious, yet sometimes soaring at the full reach of intelligence and emotion, marks Meredith's passionate search for generosity and for order. Comedy for him marked one kind of order: something bracing, tonic, but curiously sly. The later novel of his that directly calls in – and pinpoints problems in – Darwin's thinking is *The Egoist*.

This is a novel that prances along. Its brilliance and its insight are all placed at the disposal of its heroine, Clara Middleton, who finds herself trapped in the narrow confines of a courtship and imminent marriage by Sir Willoughby Patterne, gentleman, landowner, dictator, and egoist. Willoughby is introduced to us early in the book greeting his devoted Laetitia on his return from a three year absence:

> Laetitia was the first of his friends whom he met. She was crossing from field to field with a band of schoolchildren, gathering wild flowers for the morrow May-day. He sprang to the ground and seized her hand. 'Laetitia Dale!' he said. He panted. 'Your name is sweet English music! And you are well?' The anxious question permitted him to read deeply in her eyes. He found the man he sought there, squeezed him passionately, and let her go.[20]

Willoughby gets his come-uppance, but not before we have been led down the devious paths of motive and shared the panache of epigrams that carry sharp messages about human behaviour. Clara is young, intelligent, inexperienced, resistant to the demands Willoughby makes for total oneness between them, excluding the world (which she comes to realize means obliteration of herself). He seeks a promise from her that she will never marry again if he dies: she's revolted by this excess, this extravagance, of possession. In particular she is alerted by his behaviour to those dependent on him. Anyone who seeks independence is made, in Willoughby's word, 'extinct': 'He becomes to me at once as if he had never been. He is extinct' (128). Willoughby simply seeks to obliterate people, refusing any further contact. He won't re-employ a workman with nine children because he went off to try to be a shopkeeper. He threatens to 'extinguish' his cousin Vernon if he goes to live in London. Perhaps Meredith winced here at his own behaviour to Arthur, so beloved in childhood, and then distanced and kept abroad once his new marriage began. (Arthur died at the age of 37, devotedly cared for by his half-sister Edith and her husband.)

Meredith is here also having fun with Darwin's 1871 *The Descent of Man* which formulates the idea of 'sexual selection'. The span of Meredith's career took in the first impact of Darwin's ideas. He assimilated the realization of natural change and struggle and the relish for diversity that underpinned Darwin's work. But he was rightly chary about the implications of the phrase 'the survival of the fittest', originally, as pointed out earlier, Herbert Spencer's, but adopted by Darwin in late editions of the *Origin*. Meredith, like others of his time, perceived that the shift from emphasizing 'fit' – the appropriateness of organism and environment to each other – to 'fittest', seeming to imply strongest, most powerful, shifted the balance of Darwin's insights. Darwin in the *Descent*, indeed, argued that whereas in most species the female does the selection, favouring mates for their beauty or prowess, among human beings things have been perverted, so that money buys beautiful women, and men also seek out women who are rich, and therefore probably favour the outcome of low-fertility families where all the wealth is concentrated in one girl. The narrator in *The Egoist* comments wryly on the distortion of sexual selection:

> We now scientifically know that in this department of the universal struggle, success is awarded to the bettermost. You spread a handsomer tail than your fellows, you dress a finer top-knot, you pipe a newer note, have a longer stride; she reviews you in competition and selects you [...] Science thus – or it is better to say – an acquaintance with science facilitates the cultivation of aristocracy. Consequently a successful pursuit and a wresting of her from a body of competitors, tells you that you are the best man. What is more, it tells the world so. (71–2)

Whereas Darwin emphasized the kinship of all organic life and their interdependence across species and classes, past and present, society in the person of Willoughby has chosen to emphasize only the competitive elements within his theory. This obsessive competitiveness, the book teaches, will undermine itself and all true social bonds. Wisely, Meredith allows Willoughby respite and marriage with a woman grown more sharp-eyed about him than at the outset: Laetitia; while Clara makes a marriage of passion and good sense with Willoughby's cousin Vernon. Thus, Meredith preserves the vigour and the measure of comedy while rooting out in the figure of Willoughby much that is familiar to him *within himself* quite as much as in others. So in this novel a free life is achieved and the inturned extravagance of narcissism is chastened.

George Meredith ended his life as an acclaimed writer, both novelist and poet. But his path to the Order of Merit had not been smooth. Meredith seemed to many of his contemporaries to be a hyper-modern figure, asking too much of the reader and undermining many assumptions and shared preferences. A cartoon in *Punch* in 1894, late in his life, shows him as a handsome bull in a china shop, tossing vessels here and there, with one labelled 'grammar' underfoot and one breaking up in mid-air labelled 'construction'.[21] That cartoon comes in the wake of his last novel, *One of Our Conquerors*, undoubtedly a tough read, as Meredith himself acknowledged. The first scene takes a city gentleman crossing London Bridge who falls flat on his back, upended (onomatopoeically) by 'some sly strip of slipperiness', evidently fruit from the nearby markets.[22] Is it a banana skin, that staple of raucous physical comedy? – Meredith transforms the fall and the gentleman's graceless encounter with his working class rescuer into a complex exploration of half-conscious inner life as well as a meditation on political power and class conflict. The absurdity of the fall itself remains important: the slippage between comedy and overwhelming psychological analysis characterizes several of

Meredith's late works and positions him as opening the way to modernism. Indeed, T. S. Eliot's *The Waste Land* alludes obliquely to *One of Our Conquerors* in the scene on London Bridge.[23]

There is in all Meredith's work a fascination with excess and obsession as much as with ecstatic outpourings. Surrounded as he was in Surrey by deeply wooded landscapes, he heard the cry of death as well as the flutings of spring. Meredith responded to the sense of danger and of plenitude that Darwin always marked in his descriptions of Nature:

> Enter these enchanted woods,
> You who dare.[24]

In 'The Woods of Westermain' he responded also to the evolutionary idea of exceeding all that is, responding to the drive of change:

> Then you touch the nerve of Change,
> Then of Earth you have the clue;
> Then her two-sexed meanings melt
> Through you, wed the thought and felt.
> Sameness locks no scurfy pond
> Here for Custom, crazy-fond:
> Change is on the wing to bud
> Rose in brain from rose in blood.
> Wisdom throbbing shall you see
> Central in complexity.[25]

Though now less read, he is still familiar through Vaughan William's fantasia, 'The Lark Ascending', which is nearly always at the top of people's choices in classical music. Over 15 minutes the violin above the orchestra twists and soars, trills and mounts, in a spectacular imagining of the lark's free song. Behind that music lies Meredith's poem 'The Lark Ascending', its first 80 short lines all breathed in a single sentence.[26] Before human speech, Darwin suggested, lies the intricacy and extravagance of birdsong. Meredith and Vaughan Williams seek to realize the extremes of that utterance, which is so effortlessly performed by the skylark.

> He rises and begins to round,
> He drops the silver chain of sound,
> Of many links without a break,
> In chirrup, whistle, slur and shake,
> All intervolved and spreading wide,
> Like water-dimples down a tide

> Where ripple ripple overcurls
> And eddy into eddy whirls;
> A press of hurried notes that run
> So fleet they scarce are more than one,
> Yet changeingly the trills repeat
> And linger ringing while they fleet,
> Sweet to the quick o' the ear, and dear
> To her beyond the handmaid ear,
> Who sits beside our inner springs,
> Too often dry for this he brings,
> Which seems the very jet of earth
> At sight of sun, her music's mirth,
> As up he wings the spiral stair,
> A song of light, and pierces air
> With fountain ardour, fountain play,
> To reach the shining tops of day,
> And drink in everything discerned
> An ecstasy to music turned,
> Impelled by what his happy bill
> Disperses; drinking, showering still,
> Unthinking save that he may give
> His voice the outlet, there to live
> Renewed in endless notes of glee,
> So thirsty of his voice is he,
> For all to hear and all to know
> That he is joy, awake, aglow;[27]

Here Meredith leaves comedy behind, with its play of inhibition and embarrassment. He leaves behind too, for the moment, the curtailing ironies with which his novels defend the characters and the writer. The zeal, extremes and precocities of his style break up the social and psychological systems by which he was surrounded. Though sometimes daunting, sometimes exasperating, his extravagance breaks open the moulds that set emotions in expected shapes and politics in hardened hierarchies.

There are no easy genealogies to be tracked between these three writers: response, interplay, overlap, yes, but to very varying degrees. Both Tennyson and Meredith had formed strong intellectual identities before Darwin published the *Origin*. Darwin shared plangency with Tennyson and may have learnt from *In Memoriam*. Meredith admired early Tennyson but spoke slightingly of his later 'half-yards of satin'.[28] And the style of their extravagance is not all the same: abundance, recurrence, ricochet. Systems survive and are essential to meaning. But extravagance is for each of these writers a way of imagining the world at full stretch, and watching it change. The Romantic Sublime is transformed into later nineteenth-century Extravagance, an extravagance that rushes towards new forms of feeling and knowing, new ways of combining them in the wake of Darwin's 'two-sexed meanings'.

Notes

1 Benjamin Disraeli, *Tancred; or, The New Crusade* (London: Peter Davies, 1927), 112.
2 John Stuart Mill, *Autobiography*, ed. Jack Stillinger (Oxford: Oxford University Press, 1971), 89.
3 Ibid., 91–2.
4 William Wordsworth, *The Excursion*, in *Poetical Works*, ed. Ernest de Sélincourt (Oxford: Oxford University Press, 1973).
5 'Excursion': definition 3: 'Progression beyond fixed limits'. 'Excursive': 'Rambling; wandering; deviating'. Samuel Johnson, *Dictionary of The English Language*, vol. 2 (London: Longman, Hurst, Rees, Orme and Brown, 1818).
6 Charles Darwin, *On the Origin of Species* [1859], in *Works of Charles Darwin*, ed. Paul H. Barrett and R. B. Freeman (London: William Pickering, 1988), 64.
7 *Götterdämmerung* ('The Twilight of the Gods'), the last in the four-opera cycle, was written from 1861 to 1874.
8 Gustav Mahler, *Das Lied von der Erde*, Philharmonia No. 217 (Wien-London: Philharmonia Partituren, n.d.) 137–46.
9 Charles Darwin, *On the Various Contrivances by which British and Foreign Orchids Are Fertilised by Insects, and on the Good Effects of Intercrossing* (London: John Murray, 1862), 348–9.
10 Herbert Spencer, 'A Theory of Population Deduced from the General Law of Animal Fertility', in *Religion in Victorian Britain*, ed. Gerald Parsons, James R. Moore and John Wolffe, 4 vols (Manchester: Manchester University Press, 1988), 3:405–7.
11 Christopher Ricks, ed., *The Poems of Alfred Tennyson* (London: Longman, 1969), 912.
12 The longing for and pains of isolation had of course been one of Tennyson's great themes since the start of his career, witness 'The Palace of Art', in *Poems of Tennyson*, 400–418.
13 *Poems of Tennyson*, 1186.
14 Ibid., 911–12.
15 'Locksley Hall Sixty Years After', in *Poems of Tennyson*, 1365, lines 155–6.
16 Ibid., 1365–6, lines 159–74.
17 In S. M. Ellis, *George Meredith: His Life and Friends in Relation to His Works* (London: Grant Richards, 1920), 70.
18 George Meredith, *The Ordeal of Richard Feverel*, ed. Edward Mendelson (London: Penguin, 1998), 10.
19 *The Poetical Works of George Meredith*, ed. G. M. Trevelyan (London: Constable, 1912), 133.
20 George Meredith, *The Egoist* (Harmondsworth: Penguin, 1968), 59.
21 *Punch*, 18 July 1894. Reprinted in *George Meredith*, 220.
22 George Meredith, *One of Our Conquerors* (London: Archibald Constable, 1908), 1.
23 T. S. Eliot, *Selected Poems* (London: Faber, 1965), 53.
24 *Poetical Works of Meredith*, 193.
25 Ibid., 198.
26 Ibid., 221.
27 Ibid., 21, lines 23–33.
28 Ellis, *George Meredith*, 71.

Chapter 10

T. H. HUXLEY, SCIENCE AND CULTURAL AGENCY

Jeff Wallace

In a pioneering study of T. H. Huxley published in 1978, James Paradis made the claim that Huxley created, in his writing and public speaking combined, a 'unique cultural agent' – 'the scientist'.[1] In the present essay I want to explore the efficacy of this conception of science as cultural agency in terms of the encounter it stages or implies between, on the one hand, contemporary interdisciplinary protocols in the study of science within a humanities context, and, on the other, the field of debate between science and culture in Huxley's milieu and work. In addressing the cultural dimension of Huxley's intellectual work, Paradis was not the first commentator to identify a distinctly *literary* quality in Huxley's scientific writing. The concept of 'cultural agency' itself, however, derives from a more recent and precise paradigm which Adrian Desmond, whose work on Huxley I will consider alongside that of Paradis, proudly calls the 'new contextual history of science'.[2] 'Culture' in this context (more akin to Raymond Williams's expansive definition than to Arnoldian or Leavisite conceptions of excellence) is more readily associated with what became known, variously, as the science or culture 'wars' of the later twentieth century – struggles, that is, within which science and the humanities appeared to be disputing the same kind of epistemological terrain. Postmodernism, and with it the emerging confidence that humanities disciplines such as literary or cultural studies and philosophy could address themselves with legitimacy to scientific texts, concepts and debates, together constitute the principal contexts for these wars. Their *locus classicus* within academic scholarship remains the 'Sokal hoax' of 1996, pertaining to Alan Sokal's publication in the American journal *Social Text* of a bogus article on the applicability of relativity theory.[3] The acceptance for publication of this article by the editors of the journal was held to constitute decisive evidence, for Sokal, of the unsafe ground upon which many humanities intellectuals stood in their dealings with science.

With T. H. Huxley as focal point, I want to reflect on what might be at stake in the translation of such concepts as cultural agency, and their associated contemporary contexts, into later nineteenth-century debates over the public and political role of science. As the term 'agency' suggests, this becomes a study not of scientific knowledge as such but of polemic, influence, argument, debate and hegemonic struggle. Because I believe that it has strangulated interdisciplinary debate between science and the humanities since the 1960s, I have strenuously avoided using the phrase 'two cultures' to frame the discussion (although this disclaimer might itself be taken as evidence of failure in that respect). This became progressively more difficult, however, as my discussion turned more towards adversarial rhetorics, claims and counterclaims. The substance of the essay is, often, the nature of polemical debate; and to thematize this in a minor key, I want to begin in the mode of brief cautionary anecdote, concerning my own involvement in a strident exchange of views over an article, 'Zombies and Dinosaurs: The Humanities in an Age of Science', by Roger Caldwell, published in 2000.[4]

My response was to an argument that the humanities was currently in the grip of a crisis attributable to two main factors: first, the use of pseudo-scientific jargon by practitioners of literary and cultural studies, from positions of scientific ignorance; and second, the implacable progress and 'hegemony' of science in our age, making genuine discoveries of 'absolute truth' which left the arts and humanities looking distinctly expendable, with a diminished and diminishing role as the custodians of second-order 'subjective' truths. Perhaps the trigger for this response was Caldwell's attack-mode innuendo directed towards the decoy location of 'academia', accusing at the outset of his article two unfortunates – one a senior lecturer in film studies at Sydney writing on Spielberg's *Jurassic Park*, the other an author of a book on Foucault published by 'the once-respectable' Edinburgh University Press – of 'sheer effrontery of [...] performance', 'abysses of ignorance' and 'fatuity' in their writing, which for him seemed to constitute unmistakeable evidence of terminal decline in the humanities. At the same time, however, I also found I could not assent to the general propositions of the article, such as those contained in the following:

> The laws of physics allow for no exceptions. The mathematicization of nature goes all the way down, and we ourselves are included in the equations [...]
>
> The world of science is a world of facts alone – we as human beings live in a world not only of facts but of values. The world of science is an objective one, but as human beings we live subjective lives. Science can tell us about the outside of our lives; it cannot as such enter the inside of our existence. Science offers us the world at a high level of abstraction, [*sic*] we don't, however, for the most part live our lives at a high level of abstraction.[5]

Following an approach to the editor, I was given a thousand words to reply. 'No intelligent scientist', I harangued, would today be seen dead perpetuating such 'reductively simplistic' dichotomies as those of fact/value, objective/subjective, outer/inner. Not only for my interlocutor's whipping boys, 'the postmodernists', but also for scientists of the nineteenth century – I cited T. H. Huxley, as well as Darwin and Tyndall – scientific knowledge was fragile, provisional and often expressed through an abiding dialectical tension between a growing sense of epistemological mastery and excitement at the discovery of what might really be the case about the world, and a correspondingly intense recognition of the materiality of language and cultural form within which scientific knowledge must also take shape. I generously recommended a short course of reading in the best contemporary critical work in science and literature, where no 'crass' distinctions between science and the humanities obtained, and I concluded with the view that this kind of interdisciplinary liaison signalled a viable future rather than a crisis, with the humanities preserving its role in reminding us of the 'value-laden character of all knowledge'.[6] I remember a moment of unease or hesitation as I tapped out that last phrase – something distinctly off-the-peg about it, accompanied by a vague sense that I'd loosened my grip on the argument as the finishing post was in sight. But in the spirit of brevity and polemical exchange, it was left to stand.

Sure enough, this latter inkling or fear was realized when the editor allowed a response, in turn, from Caldwell, to sit next to my riposte. Wallace does not, Caldwell maintained, offer even the ghost of a rational argument: for him, 'science can only offer us' – yes! – 'value-laden knowledge.'[7] He denies that objective truth is attainable, yet admits that scientific knowledge is somehow also cumulative. How would technology work at all, if there were no distinction between fact and value? In disdaining this distinction, Wallace's view of knowledge trivializes both scientific and humanistic versions of it. Having sought to anticipate and avoid the not uncommon charge of an indiscriminate relativism in defence of how the contemporary humanities might matter for and within science, I therefore found the fatal loophole had nevertheless been unerringly spotted and pulled, and I was undone, leaping nakedly into the epistemological abyss with the rest of Caldwell's demonized postmodernists.

At the risk of explaining away a moment of rhetorical carelessness as evidence of a general intellectual tendency, I have since had cause to reflect on how the slide might occur, from the assertion of an epistemological intertwining of culture and science within certain (for example, linguistic) conditions to the deduction that this constitutes a single overarching condition – 'value-laden' – for both knowledges. Similarly, the proposition of a benign mutuality between science and the humanities as they play out their roles, 'often but not always in very different ways', in knowledge formation, might – my writing had demonstrated – glide effortlessly into the insinuation of the *priority* of the cultural as a meta-discourse which is able to 'remind' science of what it might not know – that is, that 'all' knowledge is value-laden.[8] To re-approach: it does not follow from Huxley, Darwin and Tyndall's awareness of the materiality of language – a materiality they experienced primarily as *writers* – that all knowledge may thenceforth be relativized as the product of arbitrary systems of language. My intention had been to suggest that this awareness of language emerged within their 'science', as part of it rather than as a necessary obstacle to it, and that this was not incompatible with a conception of science as a relatively autonomous field of enquiry quite reasonably founded upon the principle of objectivity.

In recent reflections upon the rise and tribulations of 'theory' as an autonomous discourse, François Cusset offers a historical framework within which such errors might at least be rationalized. Characterized by the triple negations of twentieth-century philosophy – '*antireferential, antifoundational* and *antirepresentational*' – , theory, Cusset argues, inevitably emerged into a fiercely adversarial post-war field. From an 'ever-renewed list of enemies' ranging from metaphysicians through to empiricists and functionalists, 'hateful reactions' to the threat of conceptual relativism came to be expected, and theorists could become polarized towards what Cusset calls the option of the 'discursive spiral':

> But in having to defend theory's right against its many enemies' wrongs, the new transdisciplinary discourse of theory born from the intellectual crisis of the 1960s got itself into a discursive spiral, trying to catch its moral and political enemies in the web of discursive and linguistic relativity, always on the edge of relativ*ism*, forced to overstate aporias and the deconstructive nature of language as the best retaliation against such attacks, and often losing track of the political and ethical stance which had first inspired them to reload theory, in favour of an endless regression into the all-discursive argument. This is a spiral to the extent that the new tenets of theory have often unwittingly vindicated those who had reduced their various endeavours to a late, postmodern version of the famous 'linguistic turn'.[9]

Those of us who have worked in the field of Desmond's 'new contextual history of science', taking various routes to the endorsement of the 'cultural' in scientific contexts, know what is at stake in the similarly unforgiving battles of the culture wars. In contrast to Huxley's situation, it can appear that the balance of power is freighted towards a scientific field which, as ventriloquized in Caldwell, claims to hold a monopoly both on knowledge and on analytical rigour. It is such circumstances that make it, for example, possible for a scientist of Richard Dawkins's reputation to assert that one of Britain's leading moral philosophers is 'educationally over-endowed' with the tools of her discipline – the said philosopher designated, somewhat breathtakingly, as 'someone called Mary Midgley'.[10] Dawkins then positions philosophy on a spectrum alongside the more literary orientations of the humanities, courtesy of P. B. Medawar's observation on 'the attractions of "philosophy-fiction"' to '"a large population of people, often with well-developed literary and scholarly tastes, who have been educated far beyond their capacity to undertake analytical thought."'[11] At least E. O. Wilson is able to acknowledge the danger of sounding 'patronizing' in recommending either simply 'walking away' from Michel Foucault or advising the philosopher, if it were possible, that 'it's not so bad.' Once we accept that the Foucauldian deconstruction of subjectivity simply follows from the formation of a material universe without humans 'in mind', Wilson contends, existentialist despair can be abandoned: 'Realism will be advanced to new levels, and emotions played in potentially infinite patterns. The true will be sorted from the false, and we will understand one another very well, the more quickly because we are all of the same species and possess biologically similar brains.'[12] Postmodernism can then be 'saluted' for the useful role it plays in providing relief for those who have 'chosen not to encumber themselves with a scientific education', and the rather more invaluable role of helping to strengthen organized knowledge through the need for defence against 'hostile forces'.

Nevertheless, it is scarcely viable to contest serene scientific fundamentalisms such as those of Caldwell, Dawkins and Wilson through an equally complacent, if unwitting, recourse to parallel culturalist imperialisms. If the practice of our new contextual history of science harbours its own version of culturalism, there are strong incentives to ensure that this does not generate unforeseen reconnections with the variety in which, for example for F. R. Leavis, it was only literary culture and its intimate association with the 'Third Realm' of language that could legitimately apprise science of its limits. Leavis matters here because, as I will illustrate below, it remains possible to see a full appreciation of T. H. Huxley's work as blocked by a version of British intellectual culture still powerfully informed by the priorities of Leavis and Cambridge criticism.

With these contexts in mind, it is necessary now to return to Huxley and to James Paradis on the 'unique cultural agency' of Huxley's scientist figure. As I have suggested, prominent in this definition are Huxley's literary skills. 'One of the great examples of Victorian interdisciplinary sensibility', argues Paradis, 'Huxley combined a literary sensitivity and talent with an intimate understanding of the great theoretical and practical developments in nineteenth-century British and continental science.'[13] Because literary talent implies a certain manipulative linguistic faculty – rhetoric, artifice, persuasion – the amalgamation of science and literature in Huxley also carries a political charge. Paradis's argument slides from the claim that Huxley was a scientist with literary abilities to the much more precise claim that the 'scientist figure' was Huxley's 'most significant literary creation'. In his second career as essayist and cultural critic, Paradis maintains, Huxley fused the 'voices' of critic and scientist, thereby installing this new scientist-as-cultural agent as the 'primary intelligence of his essays':

> Essentially a persona, a second self extracted from professional experiences, historical antecedents, and personal ideals, Huxley's idea of the scientist generated a unified vision which lent the essays their consistency of tone, perspective and value, or what Oliver Elton has called their 'noble unity of temper.' The scientist figure became Huxley's most significant literary creation, for it allowed him to formulate and sustain what amounted to a scientific world view – a critical consciousness that was able to range freely over diverse materials and to judge them according to ideals and standards it associated with science.[14]

It is worth pausing to reflect here on the moves in Paradis's argument. Writing is predicated as an act of impersonality, insofar as it is constituted by the manipulation and reconfiguration of conventional tropes, and because in this sense the written 'I' is not the instantiation of the speaking self, a process of distancing takes place in the critical account. The normative assumption that Huxley 'is' a scientist is displaced onto a 'second self' fusing the idea and the figure of the scientist. We must be reminded that Huxley and the scientist no longer, as it were, coincide (though in this light 'essentially a person' looks like an unsought paradox), and that Huxley cannot be said to *have* a scientific worldview; instead, he is permitted to 'formulate and sustain what amounted to' a worldview by the scientist 'he' (presumably located elsewhere) has created. The implication is that installing the rhetorical device of the scientist frees Huxley to do things other than science; the critic thus reveals Huxley's apparent ingenuity in creating – to what degree consciously or unconsciously, we do not know – a strategic device whose cultural 'agency' takes on a more conspiratorial hue, as if science, with the help of literary skill, is being used to further some more authentic or originary motive.

The strategic value of Paradis's approach, as an intervention in histories of science, is underlined when we turn to Adrian Desmond's account of Huxley. 'Historians', Desmond argues, 'have long accepted Huxley's claim that "the work of the popular expositor" was simply the conversion of "the hieratic language of the experts into the demotic vulgar tongue"'.[15] But, throughout the biography, Desmond warns against this innocent reading of Huxley's work. Literary skill is again a crucial factor in the mix: Desmond proposes that, unlike most scientists, Huxley wrote 'scintillating prose', and he excavates from a writer for the *Daily News* the priceless gem that 'Huxley would have known how to make Herbert Spencer readable.' More penetratingly, however, and in line with a poststructuralist critique of logocentrism and realism, it must be established that the issue of this scintillating prose is to render itself invisible. Repeatedly, Desmond exposes the claims of Huxley's prose to meta-language, transparency and neutrality: 'Neither Huxley nor his audience was disinterested,' he argues; beyond, or better because of, a 'seemingly see-through style', Huxley 'was refracting the light of science through an ideological lens'.[16] As in Paradis's account, it becomes difficult not to see the postmodern scare quotation marks around the word science; outside of these inverted commas, however, lies what Desmond describes as Huxley's 'real work', namely 'as a publicist – a one-man lobbying machine'. 'The last thing he was selling' – the word is used pointedly, lest we should be in any doubt about Huxley's ideological position as bourgeois scientist – 'was disinterested science. His science was instrumental, it had a political payoff that changed with the context.'[17]

To reprise, Paradis and Desmond construct persuasively complex accounts of the linguistic, cultural, and finally ideological and political determinations of science in Huxley's work. Paradis tentatively inaugurates this process, reminding us that Huxley the critic or essayist remained a second career, yet at the same time helping to create the circumstances within which the primary career of scientist could be problematized by the construction of a rhetorical 'scientist figure' through which Huxley was 'allowed' to have a scientific worldview. Desmond's more heroic narrative, driven by a sense of belonging to a 'beleaguered minority' with an argument to win – 'the onus lay on us to prove that science really was socially contingent' – completes the process through a more fully developed hermeneutic of suspicion: while Huxley said he was doing science, he was really doing cultural agency, which in its turn is really politics.[18] Again, those practising the new contextual history of science know only too well the provenance and importance of arguing the ideological case against the fundamentalists of value-free science. What might be gained, however, if in the act fixing science within its quotation marks, cultural agency and politics are floated free of them? Why de-essentialize Huxley-as-scientist, if the same move threatens to essentialize Huxley-as-cultural-agent? Are there any circumstances in which science might or should be allowed the same kind of relative autonomy, or, more to the point, any grounds upon which a new contextual history of science would predicate Huxley as a scientist without irony or the uneasy glance over the shoulder?

A third account of Huxley's intellectual work is germane to these questions. Neil Belton offers a spirited endorsement of Huxley, though on significantly different terms.[19] Why, asks Belton, don't we read Huxley today? The proposed historical answer lies in a British national culture whose deeply ingrained hostility to science is expressed in its preference for the 'irrepressible triviality' of grandson Aldous over grandfather Thomas Henry. This national culture, particularly in the form of an ideology of Englishness, is characterized, Belton argues, by a persistent stifling of the scientific 'imagination'; one of its many 'antiquated barricades' is the strange, 'neo-Gothic door that controls access from science to other kinds of thinking and making'.[20] By the mid-twentieth century, Belton continues, England had 'found its most articulate *native* voice in the field of literary criticism. Literature was elevated to be the moral arbiter of the humanities, and of human values generally'; he cites a confluence of vectors including Bloomsbury, Leavis, the left wing 'culture and society' tradition and the 'highly-ordered networks' of literary London. 'For these versions of culture', it is asserted, 'science was, it almost went without saying, an insult to human values'; figures like Huxley could have little resonance in a national culture in which so much of the 'creative energy of rationalism' had been snuffed out.[21]

As in Paradis and Desmond, Huxley's ability as a writer plays a key role in Belton's assessment: he was 'one of the most sombre and gifted polemicists in the language'. But what is simultaneously valued here is, in Burke's terms, a 'wrestling with difficulty' in plain and forceful language. It is therefore possible, in Belton's argument, to separate linguistic skill from the potential difficulties of a purely *literary* ability. He notes that, in an essay entitled 'T. H. Huxley As a Literary Man' (1935), Aldous Huxley applied to his grandfather's prose a literary criticism of 'quite stunning banality', praising his ancestor for his use of allusion, alliteration and caesura sentences.[22] Like the pigs he would make out of orange peel for his grandchildren at Christmastime, the literary qualities of his essays took on a quality of timeless or at least mythological artifice, lifting them out of the domain of mere scientific utility and ensuring the elder Huxley's abiding reputation as a 'man of letters'. Belton is dismayed not only that Huxley's work receives the patronage of his insufferable grandson but also that scientific discourse can be appropriated by a dominant literary culture and turned into empty formalism – 'as though', he bemoans, 'T. H. Huxley's earliest and most powerful defences of Darwinian theory were contentless: candied porkers in a graceful tissue of style.'[23]

In the spirit of Huxley's rhetorical tendency to invoke his forbears ('if Priestley were amongst us today […] '), let us assume that it is permissible, for a moment, to reflect upon exactly how a T. H. Huxley, suddenly transmogrified here in the early twenty-first century, might respond to his role within our recent culture wars. What in particular might the accounts of his own work in Paradis, Desmond and Belton mean to him? Assuming a certain gratitude for the generous and detailed biographical endorsements of Paradis and Desmond, it seems unlikely that this would extend to a humble acceptance of being found out as a subtle rhetorician and bourgeois ideologue all along. We are reminded that in Desmond's passionate argument, Huxley serves the needs of a contemporary cause – that of the new contextual historians of science versus a traditional history of scientific 'ideas' and, correspondingly, a relatively autonomous domain of scientific knowledge. Its overarching irony is that the virtues for which Desmond champions Huxley – those of the cultural agent of science – are in an important sense precisely the opposite of those for which Huxley wished to be valued. For Huxley, that is, science really did need to be acknowledged in and for itself, in a way which dissociated it from dominant culturalist prejudices and with them considerations of language and 'mere learning'.

Huxley's preference would surely, therefore, be for Belton's account. Looking around him, and surely with satisfaction, to find science established so decisively as a central element in school and higher education, maintaining its levels of research funding and ring-fenced from the savage evisceration of the post-Browne funding settlement, Huxley might nevertheless be intrigued to find that a critic such as Belton still found the need to highlight a set of dominant values which are literary in orientation. In 'Science and Culture' (1880), Huxley wrote:

> How often have we not been told that the study of physical science is incompetent to confer culture; that it touches none of the higher problems of life; and, what is worse, that the continual devotion to scientific studies tends to generate a narrow and bigoted belief in the applicability of scientific methods to the search after truth of all kinds? How frequently one has reason to observe that no reply to a troublesome argument tells so well as calling its author a 'mere scientific specialist'.[24]

Conveyed here is a vivid sense of the entrenched prejudices ranged against the professional and educational claims of science in late Victorian culture. Huxley goes on to point out that two of his deepest convictions about literary (classical) education – that it is of little or no direct value to the student of physical science, and it is not superior to a scientific education for the purposes of 'attaining real culture' – are 'diametrically opposed to those of the great majority of educated Englishmen, influenced as they are by school and university traditions'.[25]

The openness of this polemic, sustained across many of the essays of this period, demonstrates science's need of cultural agency in Huxley's eyes.

Recent accounts of cultural agency as an *ideological* device in Huxley therefore lose some of their revelatory quality when seen from this perspective; Huxley was perhaps more explicit and open about the cultural and ideological value of scientific objectivity and what he was doing on its behalf than it helps new contextual historians of science to assume. However scandalous, in its context, Huxley's insistence on science's equivalence to culture had been, there is a sense in which it might continue to scandalize us. Steeped in conflict with the scientific ideologues of the present, how tempting is it for us to identify Huxley with the same adversaries? Would our new culturalism easily allow the possibility — however paradoxical — that Huxley's defence of scientific autonomy might itself constitute the ground of an enlightened and progressive approach to science as culture? As I have suggested, an interdisciplinary approach to science as culture has emanated from the humanities, where a tendency to assume that only the humanities can remind science of its cultural dimension has taken hold; and I write as someone who has rather too easily committed himself in print to this position, through seductive simplifications like 'the value-laden character of all knowledge'.

Belton's suggestion is that we may have lost a sense of what it really meant for Huxley to argue on science's behalf. Where Adrian Desmond cautions us against the naïve acceptance of Huxley's version of scientific method — with the general guiding principle that it is more emancipatory to reveal science as cultural agency that it is to do science — Belton proposes that we consider anew Huxley's 'creative rationalism'. To see science as *itself* the driving force behind Huxley's cultural agency, and even behind an emancipatory politics, rather than as a product of it, we need to be able to believe — both strategically and actually, as it were — that science, to return for a moment to my own interlocutor Caldwell, can be correct about things. Before we see this as quaint and dangerously credulous, we might consider its principle of Enlightenment rationalism to be little different from our wish to expose the mystifications of science as a master narrative. Would we wish our *exposé* of Huxley's ideological science to be, itself, exposed as value-laden, or rather would we prefer that it were simply right? Huxley himself, of course, insisted in similar fashion that science was no more than an extension of everyday, progressive rationalism: it was, he wrote, 'simply the mode at which all phenomena are reasoned about, rendered precise and exact', '*nothing but trained and organised common sense*', its results won by 'no mystical faculties, by no mental processes, other than those which are practised by every one of us, in the humblest and meanest affairs of life'.[26] What might follow from consenting to this on epistemological grounds before, or alongside, identifying its potential for the inculcation of false consciousness?

In examining the relationship between science and politics in Huxley's cultural agency, across a range of his writings from the 1860s through to the publication of 'Evolution and Ethics' in 1893, what we find is a mobility of position and argument directly relating to his conviction of 'the utterly conditional nature of all our knowledge' and to 'the danger of neglecting the process of verification under any circumstances'.[27] The very openness of science is, or should be, the basis of its influence as a medium of cultural democracy. Such politics needs, of course, to be heavily qualified, accompanied as it is by a sense of national economic polity which Adrian Desmond, in his earlier study of Huxley and palaeontology in London between the years 1850 and 1875, characterized as instrumentally concerned with '*the stabilisation of capitalist society*'.[28] Huxley undoubtedly saw scientific education as the necessary basis of both national social stability and international commercial competitiveness. Both are shaped by the Darwinian struggle for existence: labour must be kept cheap, and care taken not to overeducate those in whom a technical competence and functionality represents a maximum potential contribution to the health of the social organism. Yet in 'Science and Culture', it is made plain that if science has faced the considerable opposition of a dominant literary culture, so equally has it faced the opposition of the emergent industrial-capitalist bourgeoisie. Businessmen believed that their particular idol, practicality or rule of thumb, had been the source of 'past prosperity' and would 'suffice for the future welfare of the arts and manufactures'. Huxley lamented, in this commercial context, the invention of the phrase 'applied science', arguing that what passes for the latter is always the application of 'pure science' to specific problems.[29]

Neither is Huxley's science in any easy sense an ideology for the statesman, as his later interventions in debates around social Darwinism make clear. Consistently, whilst adhering to Malthusian demographics and wary of the perils faced by societies which sought to remove checks on reproduction, Huxley nevertheless disdained the readily available step towards eugenic selectionism. In the final essays of his life, 'Evolution and Ethics' and 'The Struggle for Existence in Human Societies', and through a coded lexicon for the eugenicist tendency – 'pigeon-fanciers polity', 'fanatical individualism', or 'reasoned savagery', for example – Huxley maintained that there was 'no hope that human beings will ever possess enough intelligence to select the fittest'.[30] On the contrary, those who aspired to such policies should recall those one or two occasions, common to all, when 'it would have been only too easy to qualify for a place among the "unfit"'.[31] Scientific method had brought Huxley to the point of a sustained reassessment of the relationship between human cultures and the natural or 'cosmic' process. From this position, science as cultural agency could take no other form than an undermining of any ideological function claiming to base itself on parallels between the laws of nature and the production of human cultures. Revised here are the views, expressed in 'Science and Culture', that 'social phenomena are as much the expression of natural laws as any others,' and that the nascent discipline of sociology was simply a branch of natural science that overworked biologists did not have time to develop.[32] Instead, society now is 'usefully considered as distinct from nature', it being merely a piece of sophistry or wordplay – the kind to be expected, perhaps, of a grandfather making candied porkers – that because humans are a material part of the cosmic process, so too, on the same terms, are their cultures. Really, Huxley indicates, as does any garden gone to seed, the ethical process and the cosmic process are ultimately at odds; and for the sake of 'the great work of helping one another', there is no doubt which side he is on.[33]

Any discussion of Huxley's science as cultural agency is finally obliged to consider perhaps the most conspicuous instances of this work: his addresses to working men. In these very popular lectures, Huxley's powers of rhetoric appear at their most overtly manipulative. Sustained exercises in conspiratorial innuendo, the lectures forge alliances between professional science and audiences of self-improving handicraftsmen by uniting them, in the domain of tangible facts, against a common enemy of the dominant literary culture, the 'so-called learned folks'. Addressing the Working Men's Club and Institute Union in 1877, Huxley humbly justified his presence there: 'The fact is, I am, and have been, any time this thirty years, a man who works with his hands – a handicraftsman.'[34] This, he insists, is no mere literary figure: 'I really mean my words to be taken in their direct, literal and straightforward sense'; and he is emboldened enough to suggest proving it by inviting a watchmaker into his workshop, where his guest would be set the task of dissecting the nerves of a black beetle while he himself would put together a watch (guess who would finish first?). The lecture contains one of Huxley's most barbed attacks on our 'learned brethren', whose work, he sneers, is 'untrammelled by anything "base and mechanical"'. In this heartily back-slapping, you-and-me-lads-together approach, it is language itself that is finally identified as the potential foe to be stalked out, wherever possible, by the hearty experientialism and practicality of the scientist and the worker: 'You feel and we feel that, among the so-called learned folks, we alone are brought into contact with tangible facts in the way that you are [...] You know that clever talk touching joinery will not make a chair [...] Mother nature is serenely obdurate to honeyed words [...] '.[35]

A familiar problematic re-emerges. Huxley's recurrent theme, as expressed in these lectures, is the threat of 'mere learning' or 'paper-philosophy' to the study of science and in the educational systems of late Victorian culture. As literary and cultural critics, as well as new contextual historians of science, we draw attention to the ironies of this position in a writer widely renowned for his skills of rhetorical manipulation. Further informed by a hard-won culturalist critique of scientific ideology, we note how Huxley's rhetorical skill lies precisely in the apparent denial of the materiality of language and the positing of a domain of pure practical scientific truth which lies beyond the seductions of 'honeyed words'. The question is whether this leaves us where we want to be, and how far an alternative to the temptation towards the position of linguistic analysis is available to us – one, that is, which neither simply replicates, at one end of a spectrum, the blunt yet stagey denials of the lectures to working men nor, at the other, condemns us in François Cusset's words to the 'discursive spiral'. While we examine Huxley's construction of the scientist as cultural agent, and even of himself as a kind of fellow base mechanical, what difference should it make to acknowledge that Huxley was a trained anatomist and could undoubtedly have made a fine job of the dissection of the black beetle's nerves, and even of the construction of the watch? The prompt offered by Neil Belton's argument on Huxley and knowledge comes again to mind: if we are to avoid the trap of designating Huxley's writing as 'contentless' formalism, how far do we need to be aware that the new contextual history of science, and the reconfiguration of science as cultural agency, might contribute its own means of emptying Huxley's science of content? In a fine essay on the sophisticated reflexivity of a group of key Victorian scientists, Donald Benson alerts us to the irony that it was humanists such as Arnold and Newman who 'accepted', and in this sense perhaps helped to sustain, a 'popular reductive conception' of science.[36] In the current essay I have tried to suggest how, depending on our degree of vigilance, such reductive conceptions of science might leak back into our interventions in the culture wars when we might least expect it, and when our distance from the relatively complacent humanisms of Arnold and Newman might seem most secure. The last thing we'd want, after all, would be to have Huxley, as he makes his way back to the late Victorian era, thanking us for our kind interest in his work, but regretting that we haven't quite broken free of the perils of paper-philosophy.

Notes

1 James Paradis, *T. H. Huxley: Man's Place in Nature* (Lincoln and London: University of Nebraska Press, 1978), 6.
2 Adrian Desmond, *Huxley: From Devil's Disciple to Evolution's High Priest* (Harmondsworth: Penguin, 1998), xvi.
3 Alan D. Sokal, 'Transgressing the Boundaries: Towards a Transformative Hermeneutics of Quantum Gravity', *Social Text* 46–7 (Spring/Summer 1996): 217–52.
4 See Roger Caldwell, 'Zombies and Dinosaurs: The Humanities in an Age of Science', *Planet: The Welsh Internationalist* 137 (October/November 1999): 73–82; and Jeff Wallace,

'Zombies and Dinosaurs', *Planet: The Welsh Internationalist* 139 (February/March 2000): 93–4, and Caldwell's response in the same number, 94–5.
5 Caldwell, 'Zombies and Dinosaurs', 75, 78.
6 Wallace, 'Zombies and Dinosaurs', 94.
7 'Roger Caldwell Responds', *Planet* 139, 94–5.
8 Wallace, 'Zombies and Dinosaurs', 94.
9 François Cusset, 'Theory: Madness of', *Radical Philosophy* 167 (May/June 2011): 24–30; 27 (emphasis in the original).
10 Richard Dawkins, *The Selfish Gene*, 30th anniversary edition (Oxford: Oxford University Press, 2006), 278.
11 Ibid.
12 E. O. Wilson, *Consilience: The Unity of Knowledge* (London: Abacus, 1999), 45–6.
13 Paradis, *T. H. Huxley*, 2.
14 Ibid., 6.
15 Desmond, *Huxley*, 636.
16 Ibid., 636–7.
17 Ibid., 638.
18 Ibid., 618.
19 Neil Belton, 'Candied Porkers: British Scorn of the Scientific', in *Cultural Babbage: Technology, Time and Invention*, ed. Francis Spufford and Jenny Uglow (London: Faber and Faber, 1996), 240–65.
20 Ibid., 240.
21 Ibid., 260.
22 Ibid., 242.
23 Ibid., 242–3.
24 T. H. Huxley, 'Science as Culture', in *Science and Education*, vol. 3 of *Collected Essays of T. H. Huxley* [1893–94], 9 vols (Bristol: Thoemmes Press, 2001), 140–41.
25 Ibid., 141.
26 T. H. Huxley, 'Man's Place in Nature' (1863), in *Man's Place in Nature and Other Essays* (London: Dent, 1906), 189; T. H. Huxley, 'On the Educational Value of the Natural History Sciences', in *Science and Education*, 45.
27 Huxley, *Science and Education*, 55.
28 Adrian Desmond, *Archetypes and Ancestors: Palaeontology in Victorian London 1850–1875* (Chicago: University of Chicago Press, 1982), 160.
29 Huxley, 'Science as Culture', 137.
30 T. H. Huxley, 'Evolution and Ethics' (1893), in *Evolution and Ethics and Other Essays*, vol. 9 of *Collected Essays of T. H. Huxley*, 23, 82, 115, 34.
31 Ibid., 39.
32 Huxley, 'Science as Culture', 158.
33 Huxley, *Evolution and Ethics and Other Essays*, 203, 70.
34 T. H. Huxley, 'Technical Education', in *Science and Education*, 406.
35 Huxley, *Science and Education*, 406–8.
36 Donald Benson, 'Facts and Constructs: Victorian Humanists and Scientific Theorists on Scientific Knowledge', in *Victorian Science and Victorian Values: Literary Perspectives*, ed. James Paradis and Thomas Postlewait (New Brunswick, NJ: Rutgers University Press, 1985), 299–318.

NOTES ON CONTRIBUTORS

GILLIAN BEER is King Edward VII Professor Emeritus of English Literature and former president of Clare Hall, University of Cambridge. She was made a Dame in 1998, is a fellow of the British Academy and a Foreign Honorary Member of the American Academy of Arts and Sciences. She began her career studying and theorizing Romance, in *Meredith: A Change of Masks* (1970) but moved on to work on the interpenetration of Victorian literature and Victorian science, notably in *Darwin's Plots: Evolutionary Narrative in Darwin, George Eliot and Nineteenth-Century Fiction* (1983; 3rd ed. 2009) and in *Open Fields: Science in Cultural Encounter* (1996). She is general editor of the monograph series, 'Cambridge Studies in Nineteenth-Century Literature and Culture', has published a series of essays on rhyme and is currently completing a study of Lewis Carroll's *Alice* books.

GOWAN DAWSON is a senior lecturer in the Victorian Studies Centre at the University of Leicester, where he is a Leverhulme Trust Research Fellow for 2012–13. He is the author of *Darwin, Literature and Victorian Respectability* (Cambridge University Press, 2007) and co-author of *Science in the Nineteenth-Century Periodical: Reading the Magazine of Nature* (Cambridge University Press, 2004). He is completing a new book entitled *Show Me the Bone: Reconstructing Prehistoric Monsters in Nineteenth-Century Britain and America 1795–1910*, as well as an edited collection with Bernard Lightman, called *Victorian Scientific Naturalism: Community, Identity, Continuity* (University of Chicago Press, forthcoming 2014).

ROGER EBBATSON is currently a visiting professor at Lancaster University, having previously held appointments at the University of Sokoto, Nigeria, the University of Worcester and Loughborough University. His publications include *Lawrence and the Nature Tradition* (1980), *The Evolutionary Self* (1982), *Hardy: Margin of the Unexpressed* (1993), *An Imaginary England* (2005) and *Heidegger's Bicycle* (2006). His most recent book, *Landscape and Literature 1830–1914* (2013), explores representations of nature through the frame of the Frankfurt School concept of the 'aura'. He is a member of the publications boards of the Tennyson Society and the Hardy Society and of the executive committee of the Jefferies Society.

GEORGE LEVINE is a professor emeritus of Rutgers University and was one of the founders of the literary journal *Victorian Studies*. He published a series of influential studies of Victorian prose, but his interests have been consistently interdisciplinary, as in the edited collection *One Culture: Essays in Science and Literature* (1981) and he has increasingly focused on the work of Charles Darwin and its impact on nineteenth-century culture in books such as *Darwin Loves You: Natural Selection and the Re-enchantment of the World* (2009), *Darwin and the Novelists* (1988), *Dying to Know: Scientific Epistemology and Narrative in Victorian England* (2002) and *Realism, Ethics and Secularism: Essays on Victorian Literature and Science* (Cambridge University Press, 2008), which won the British Society for Literature and Science's annual book prize. His latest book is *Darwin the Writer* (Oxford University Press, 2011).

MICHIEL NYS recently completed his doctorate at the University of Leuven and the Fund for Scientific Research, Flanders (FWO), and is a founding member of the Leuven Interdisciplinary Platform for the Study of the Sciences. He read Germanic languages at the University of Leuven and Literary Studies at the University of Leuven and the University of British Columbia, Vancouver. His research project focuses on the history and the rhetoric of popular science, and interactions between evolutionary biology, creative literature and cultural criticism, through a comparative case study of work by Thomas Henry Huxley and Julian Huxley.

VALERIE PURTON is a professor of Victorian literature at Anglia Ruskin University in Cambridge. She is the author of *Dickens and the Sentimental Tradition: Fielding, Richardson, Sterne, Goldsmith, Sheridan, Lamb* (2012) and is the co-author, with Norman Page, of *The Tennyson Literary Dictionary* for Palgrave Macmillan (2010) and, with Christopher Sturman, of *Poems By Two Brothers: The Poetry of Tennyson's Father and Uncle* (1993). She has published many articles on Victorian literature and is the editor of the *Tennyson Research Bulletin* and a member of the Tennyson Society executive committee and publications board.

MATTHEW ROWLINSON is a professor in the English department and the Centre for Theory and Criticism at Western University in London, Ontario and before that was associate professor at Dartmouth College. He is the author of *Real Money and Romanticism* (Cambridge University Press, 2010) and *Tennyson's Fixations: Psychoanalysis and the Topics of the Early Poetry* (University of Virginia Press, 1994) as well as articles and reviews on literature of the

Victorian and Romantic periods. His current projects include a book on symptoms, taxonomy and temporality in the Victorian era and a teaching edition of *In Memoriam*.

REBECCA STOTT is a professor of literature and creative writing at the University of East Anglia. For the last few years her work has been focused on the interface between literature and science. She is the author of an epic study of Charles Darwin's predecessors, *Darwin's Ghosts* (2012), a biography of Darwin, *Darwin and the Barnacle* (2003) and two historical thrillers, *The Coral Thief* (2009) and *Ghostwalk* (2007). The latter involves Sir Isaac Newton and seventeenth-century Cambridge, and the former is set in early nineteenth-century Paris and concerns, among other things, the early evolutionary theory of Georges Cuvier. She has begun a third novel set in contemporary London.

JEFF WALLACE is a professor of English at Cardiff Metropolitan University, where he teaches modern and contemporary literature and leads the Arts and Humanities Research Group. His research interests include science and literature from Darwin to the contemporary, D. H. Lawrence, and the dialogue between humanism and posthumanism. He is the author of *Beginning Modernism* (2011) and *D. H. Lawrence: Science and the Posthuman* (2005), and the co-editor of volumes on Darwin's *Origin of Species*, Raymond Williams and Gothic Modernisms. He is currently researching a history of the concept of abstraction since modernism, and writing essays on Haruki Murakami and John Berger.

CLIVE WILMER is a poet and lecturer. He lives in Cambridge, where he is an emeritus fellow of Sidney Sussex College and an honorary fellow of Anglia Ruskin University. His published work includes seven volumes of his own poetry the latest of which is *New and Collected Poems* (2012). With George Gömöri he has translated the Hungarian poets Miklós Radnóti, Györgi Petri and János Pilinszky. He has edited a Penguin edition of John Ruskin's *Unto This Last and Other Writings* and is the Master of the Guild of St George, founded by Ruskin in 1871.

www.ingramcontent.com/pod-product-compliance
Lightning Source LLC
Chambersburg PA
CBHW021829300426
44114CB00009BA/385